To Breathe Freely

Maryland Studies in Public Philosophy

Series Editor: The Director of
The Center for Philosophy and Public Policy
University of Maryland, College Park

Also in this series

Income Support: Conceptual and Policy Issues
Edited by Peter G. Brown, Conrad Johnson, and Paul Vernier

Boundaries: National Autonomy and Its Limits
Edited by Peter G. Brown and Henry Shue

The Border That Joins: Mexican Migrants and U.S. Responsibility
Edited by Peter G. Brown and Henry Shue

Energy and the Future
Edited by Douglas MacLean and Peter G. Brown

Conscripts and Volunteers: Military Requirements, Social Justice, and the All-Volunteer Force
Edited by Robert K. Fullinwider

The Good Lawyer: Lawyers' Roles and Lawyers' Ethics
Edited by David Luban

Liberalism Reconsidered
Edited by Douglas MacLean and Claudia Mills

The Security Gamble: Deterrence Dilemmas in the Nuclear Age
Edited by Douglas MacLean

To Breathe Freely

Risk, Consent, and Air

Edited by
MARY GIBSON

Rutgers University

ROWMAN & ALLANHELD
PUBLISHERS

ROWMAN & ALLANHELD

Published in the United States of America in 1985
by Rowman & Allanheld, Publishers
(a division of Littlefield, Adams & Company)
81 Adams Drive, Totowa, New Jersey 07512

Copyright © 1985 by Rowman & Allanheld

All rights reserved. No part of this publication
may be reproduced, stored in a retrieval system, or
transmitted in any form or by any means, electronic,
mechanical, photocopying, recording, or otherwise,
without the prior permission of the publisher.

Library of Congress Cataloging in Publication Data
Main entry under title:

To breathe freely.

Includes index.
1. Air—Pollution—Addresses, essays, lectures.
2. Air—Pollution—Standards—Addresses, essays, lectures.
3. Air—Pollution, Indoor—Addresses, essays, lectures.
4. Air—Pollution, Indoor—Standards—Addresses, essays,
lectures. 5. Air quality—Standards—Addresses, essays,
lectures. I. Gibson, Mary.
RA576.T6 1985   363.7'392   84-22278
ISBN 0-8476-7416-8

85 86 87 / 0 9 8 7 6 5 4 3 2 1
Printed in the United States of America

Table of Contents

Tables and Figures   vii

Preface   ix

Introduction   xi

PART I   The Risks of Airborne Contaminants

1   Risks of Passive Smoking   *James L. Repace*   3

2   Airborne Contaminants in the Workplace   *Laura Punnett*   31

3   The Risks from Transported Air Pollutants   *Robert M. Friedman*   52

PART II   The Meaning and Role of Consent

4   The Role of Consent in the Legitimation of Risky Activity
     Samuel Scheffler   75

5.   Locke, Stock, and Peril: Natural Property Rights,
     Pollution, and Risk   *Peter Railton*   89

6   Imposing Risks   *Judith Jarvis Thomson*   124

7   Consent and Autonomy   *Mary Gibson*   141

8   Cognitive and Institutional Barriers to "Informed Consent"
     Baruch Fischhoff   169

9 The Role of Consent in Managing Airborne Health
   Risks to Workers   *Michael Baram*   186

PART III Pollution and Policy

10 Reform of Occupational Safety and Health Policy
    Mark MacCarthy   201

11 The Confusion of Goals and Instruments: The Explicit
    Consideration of Cost in Setting National Ambient
    Air Quality Standards   *George Eads*   222

12 Achieving Air Pollution Goals in Three Different Settings
    Clifford S. Russell   233

Epilogue: Hazards of Entering the Risk Debate

13 On Not Hitting the Tar-Baby: Risk
    Assessment and Conservatism   *Langdon Winner*   269

Index   285

Notes on Contributors   293

Tables and Figures

Tables

1.1 Comparison of Estimated Lung Cancer Mortality from Passive Smoking with Statistical Estimates of the Frequency of Death per 10^8 Residents per Year, 1968–1973 Average — 21

1.2 Boundaries of Acceptable Risk — 22

2.1 Estimates of Numbers of People Occupationally Exposed to Selected Chemicals and Other Airborne Substances — 40

12.1 Difference in Context: Three Air Quality Problems — 241

Plates

1.1 NEA cartoon — 4

1.2 Blondie cartoon — 17

1.3 Herblock cartoon — 24

1.4 Relative Indoor/Outdoor Air Pollution Levels — 25

Preface

The Center for Philosophy and Public Policy was established in 1976 at the University of Maryland in College Park to conduct research into the values and concepts that underlie public policy. Most other research into public policy is empirical: it assesses costs, describes constituencies, and makes predictions. The Center's research is conceptual and normative. It investigates the structure of arguments and the nature of values relevant to the formation, justification, and criticism of public policy. The results of its research are disseminated through workshops, conferences, teaching materials, the Center's newsletter, and books like this one. The Center enjoys the general support of the Rockefeller Brothers Fund.

This is the 9th volume of the Maryland Studies in Public Philosophy. Previous volumes, listed across from the title page, dealt with the welfare system, the significance of national boundaries, immigration, energy policy, military manpower policies, lawyers' ethics, the theoretical and practical viability of liberalism, and nuclear deterrence.

This book is one of a pair issuing from a research project funded by the National Science Foundation's program on Technology Assessment and Risk Analysis. The companion volume, *Values at Risk*, was edited by Douglas MacLean, who was principal investigator for this project. Each of these volumes grew out of a series of meetings of a working group comprising the contributors to the volume. The views expressed by the individual contributors are, of course, their own and not necessarily those of the Center or its sources of support, the National Science Foundation, or the institutions and agencies for which the contributors work. I want to take this opportunity to thank publicly the members of the Working Group on Risk, Consent, and Air not only for their hard work, but for the spirit of inquiry, cooperation, and mutual respect they brought to every aspect of this project. Thanks also to Doug MacLean, who conceived the overall project, and to Henry Shue, who was Director of the Center, for their active support throughout.

The work of administering the working group and of turning the products of its efforts into a book was shared with Elizabeth Cahoon, Louise Collins, Carroll Linkins, Lorrine Owen, and Robin Sheets. Special thanks are due to Rachel Sailer for arranging the working group sessions and being there to ensure that they ran smoothly. Finally, I want to thank Claudia Mills, the Center's editor, for the philosophical and editorial contributions she made every step of the way.

MG

Introduction

If we were confident that we would not be exposed to grave or unreasonable risks without our consent, we would, as the expression goes, breathe more freely. Wherever we go, and whatever else we do or care about in life, we all must breathe. So there is almost no way that we can avoid certain sorts of airborne risks imposed on us by others. How can we determine what these risks are? How and why does it matter how freely we bear them?

Scarcely any ordinary activity can be carried on that does not impose some risk on others. Clearly, we are not morally required to obtain the explicit consent of everyone potentially at some risk from an act before we may perform it. You need not obtain the permission of everyone who might be in the path of your car should it go out of control on your way to the post office before you make the trip. Nor must you ask the consent of your neighbors before you install electrical wiring or a heating system in your home, or make a fire in the wood stove, although the risk to them of fire is increased somewhat by these acts of yours. Equally clearly, there are some risks that it is not permissible to impose without the consent of those potentially affected. Medical researchers or practitioners may not expose you to the risk of experimental drugs or surgery without your explicit informed consent. I need your permission before I may practice my knife-throwing act on you. But it is not clear that we have a defensible way of distinguishing those cases in which consent is required from those where it is not. Further, in cases where some form of consent does seem to be required, it often is not clear which of several possible forms of consent or surrogates for consent would suffice.

This book is one of a pair resulting from an interdisciplinary project designed to explore the relations between risk and consent in general and, in particular, to address the question, "What is the proper role for individual consent in centralized policy decisions concerning risk?" The other half of the project examined these issues in more general or abstract terms, while the half represented by this book focused its discussion by concentrating on one particular category of risks: those posed by air pollution. As this division of labor worked out, the companion volume, *Values at Risk,* edited by Douglas MacLean, concentrates more on issues related to the meaning and assessment of risk in general, while the present volume focuses more, though not exclusively, on examining the meaning and role of consent. In addition, this volume explores the nature and magnitude of the specific risks associated with air pollution and critically examines specific policy proposals for coping with such risks.

We chose three different, though overlapping, arenas in which to explore airborne risks and responses to them: cigarette smoke in enclosed places, workplace air pollution, and outdoor ambient air—especially long-range transportable pollutants. We chose these particular arenas for providing both continuity and contrast among the issues to be examined. In addition to the evident contrasts, there are significant overlaps and interactions among these arenas. Airplanes, elevators, restaurants, and homes are enclosed places where persons may be exposed to cigarette smoke; they are also workplaces—homes, of course, being workplaces for both (albeit often poorly) paid and unpaid workers. Cigarette smoke contributes to air pollution in many offices and factories. The combination of smoking with other airborne hazards can increase health risks far beyond the sum of the risks associated with each hazard individually. These synergystic effects may carry over to family members if workers carry hazardous substances home on their clothing. And outdoor air poses workplace air pollution hazards to many workers—to bus, truck, and taxi drivers, police, and agricultural workers, among others.

This book is divided into three sections plus an epilogue. Each of the three chapters in the first section details what is known—as well as what is not known and in some cases perhaps cannot be known—about the risks of airborne contaminants in one of our three arenas. James Repace discusses the risks of secondhand cigarette smoke, Laura Punnett details the risks posed by airborne contaminants in the workplace, and Robert Friedman probes the risks associated with outdoor air pollutants, particularly those that travel some distance in the atmosphere. Thus, in addition to disseminating the authors' research findings and personal views on the issues, these chapters provide the reader with essential background information for the discussions in the sections that follow.

The chapters in Section II are largely devoted to examining issues concerning consent. Samuel Scheffler's essay develops and assesses three possible answers to the question "What are the underpinnings in moral theory of the moral significance of consent?" According to one view, consent has instrumental value: a social system in which consent requirements are respected will result in better decisions and happier people. Another view is that consent has intrinsic moral value as an essential component of a good life. The third view is that the importance of consent derives from its role within a system of natural rights.

Peter Railton's chapter gives a detailed exposition and critique of one of the theoretical approaches outlined by Scheffler, that of Lockean natural rights theories. Railton argues, contrary to a widespread assumption, that Lockean theory entails an extremely restrictive governmental policy prohibiting the imposition of many kinds of risks in the absence of explicit prior consent by each individual potentially at risk.

Mary Gibson attempts to develop and defend an account of the moral importance of consent in terms of its relation to autonomy. As the value underlying consent, she argues, direct appeal to autonomy can provide guidance for policy decisions in circumstances in which individual consent itself is not feasible.

Judith Thomson unravels the almost dizzying web of issues involved in attempting to say which unconsented-to risks it is permissible to impose and which one ought not impose—and what it means to say these things. In the final section of her chapter, she calls attention to a crucial and much-ignored question about consent: What precisely is the content of consent? That is, on a given occasion, just what is—and what is not—consented to? By consenting to work at a job where one is exposed to airborne toxins, for example, does one thereby consent either to the job's being risky or to the health risks themselves? By purchasing a home in a polluted area, does one thereby consent either to the pollution or to the health risks it poses?

Baruch Fischhoff examines the question of what constitutes informed consent from the perspective of cognitive psychology's behavioral decision theory. He goes on to probe the limits of this approach, considering the inadequacies of using the idea of optimal decision making as a criterion of informed consent, and of the concept of informed consent itself.

Michael Baram locates consent in the context of contract and tort law and of our societal system of values to set the stage for an exploration of the role of consent in managing health risks to workers. Surveying in turn the common law, workers' compensation law, and federal regulation under the Occupational Safety and Health Act, he considers such questions as how consent is provided for, how voluntary and informed it is, and how useful consent is within that framework for managing occupational health risks. Baram's chapter, by exploring the role of consent in currently existing policy, provides a bridge between the present section, devoted primarily to consent, and the third section, whose primary focus is on assessment of specific policy proposals.

Current occupational safety and health policy in this country is criticized for both inefficiency and ineffectiveness. Many of these criticisms are, Mark MacCarthy argues, well founded. He develops and discusses the merits and potential weaknesses of four proposals for remedying some of the system's ills. Heavier reliance on the court system is found to be initially appealing, but unsatisfactory on closer examination. More promising in MacCarthy's view are (1) incorporating worker health and safety protection into a coherent, overall industrial policy, (2) greater involvement of workers and managers in the regulatory process at the national and local levels, and (3) greater use of the collective bargaining process for enforcement of federal standards and the establishment of worker-management safety and health committees.

Should the Clean Air Act be amended to permit (or even require) the explicit consideration of costs in setting National Ambient Air Quality Standards (NAAQS)? George Eads critically explores two reasons for opposing this suggestion. One is based on the view that clean air is a right that ought not be subject to cost-benefit trade-offs. The second is based on a strategic argument to the effect that explicit consideration of costs would lead in practice to less stringent standards and hence to less adequate public health protection. Eads argues that the first view depends on an untenable dichotomy between the standards, conceived (inaccurately) as goals, and plans for implementing them, in which costs do play an explicit role. The strategic view has some merits, he concludes, but it entails costs of its own that ought not to be ignored.

Clifford Russell's contribution begins by cataloguing the many difficulties inherent in choosing among different systems for implementing air pollution control goals. Out of this account, Russell develops a list of five criteria for the assessment of proposed implementation schemes. He then systematically applies these criteria to analyze the strengths and weaknesses of both "economic incentive" and "command and control" systems in each of our three arenas: outdoor air, the workplace, and side-stream cigarette smoke.

In the Epilogue, Langdon Winner steps back and reflects on the political implications of the process in which this project has involved us all. He warns of what he sees as the hazards of entering the "risk" debate. He argues that a politically conservative bias is inherent in the terminology and the assumptions of risk assessment, and that the language of risk misdescribes some important problems and renders others virtually invisible.

Although a good deal has been written on risk, consent, and government policy concerning air pollution, our society is still a very long way from a satisfactory resolution of the issues surrounding these topics. This is the first effort of which we are aware to treat these three subjects in combination. For these reasons, it seemed appropriate to encourage the most freewheeling possible exploration of the subject matter. Our purpose was more to raise and illuminate questions than to provide tidy solutions. Indeed, it is fair to say that the need to critically examine apparently tidy solutions to large, complex, and messy problems was a significant factor motivating this entire project.

Each author has his or her specific area of expertise and his or her personal perspective both on what the important issues are and how most fruitfully to approach them. Each, therefore, was encouraged to, and does, speak in his or her own voice. We began, for example, with no assumption that there were available adequate definitions of the basic concepts with which we were concerned. Thus we did not adopt a standard definition of 'risk' or 'consent' to be employed throughout the book. Instead, we asked that each author be as clear and explicit as possible about his or her own usage, especially where different usages might affect the argument. How significant the differences are among different usages is one of many questions that remain to be explored further. It is hoped that the essays contained here, making explicit as they do important implications of the usages they employ, will contribute to that inquiry.

Another consequence of having each author speak in his or her own voice is that the reader must be left to draw his or her own conclusions. There is no single perspective presented nor set of conclusions or lessons to be drawn from a collection such as this. Each chapter speaks for itself.

PART ONE

The Risks of Airborne Contaminants

1

Risks of Passive Smoking

James L. Repace

Introduction

Smoking of tobacco before World War I was largely a male ritual involving cigars and pipes and confined in public to saloons, smoking cars, and sporting establishments, and at home, to a separate smoking parlor.[1] World War I recruited a new generation of male smokers: the War Industries Board estimated that Allied soldiers consumed 60 to 70 percent more tobacco than in civilian life. With the advent of cigarettes (Camel Cigarettes were introduced in 1913, and Lucky Strike and Chesterfield followed in the next few years), smoking gradually became generally acceptable, and the advertising campaigns of the 1920s and 1930s aimed at women made the habit not only acceptable, but chic. By 1922, New York women were smoking openly on the streets, and in 1934, Eleanor Roosevelt smoked cigarettes publicly. The old customs were discarded, until few smokers felt the need to ask permission before lighting up and both men and women began to smoke everywhere. In the United States, the habit grew steadily from a per capita smoking rate of 1,285 in 1925 to 1,450 in 1935. Consumption among men accelerated during World War II; in 1944, more than 25 percent of U.S. cigarette production was distributed to overseas forces, typically for free or at a low cost. Women's smoking rates also accelerated during the war, as millions of women entered the labor force. Thus, by 1950, per capita consumption of cigarettes reached 3,250 and attained an historic high of 4,336 in 1963. Cigarettes came to be identified as an essential social crutch. By 1966, 52 percent of men and 34 percent of women smoked cigarettes. In the face of the widespread popularity of smoking, it became virtually impossible for the nonsmoking majority of the population to avoid exposure to tobacco smoke. The code of the smoker had become one of discourtesy: "Light up and enjoy it, and to hell with those who find it objectionable."[2] At public gatherings, in business meetings, lodge halls, and theaters, at restaurants, nightclubs, and sports events, in college classrooms, locker rooms, on buses, trains, and in aircraft, in taxis, carpools, at social gatherings, and even in most homes, smoking was taken for granted.

Smoking generally became established through social pressure and the imitation of peers or role models who smoked. Once the habit was acquired, addictive processes contributed to its maintenance: the average pack-and-a-half-a-day smoker got more than 100,000 nicotine "shots" per

"It's equal rights—I have a right to smoke and you have a right to wear a mask!"

Figure 1.1
Reprinted by permission of Newspaper Enterprise Association.

year—a frequency unmatched by any other form of drug taking. Smoking maintenance was characterized by steady state behavior, fueled by nicotine withdrawal, and stimulated by alcohol consumption.[3] Currently, about 38 percent of men and 30 percent of women are active cigarette smokers (about one-third of adults) who consume two cigarettes per hour over a 16-hour waking day.[4] It appears that very few smokers can satisfy their addiction on fewer than 10 to 12 cigarettes per day.[5]

Although as long ago as 1920 there had been suggestions that cigarette smoking could cause human disease, it was not until 1964 that the U.S. Surgeon General's Advisory Committee on Smoking and Health concluded: "Cigarette smoking is a health hazard of sufficient importance in the U.S. to warrant appropriate remedial action."[6] By 1979, the scientific evidence on the health hazards of cigarette smoking was judged to be "overwhelming," with smoking judged "the largest preventable cause of death in America," causing an estimated 346,000 excess deaths per year from lung cancer and other cancers, cardiovascular disease, and chronic pulmonary disease.[7]

Effects of Tobacco Smoke on Nonsmokers

Our concern in this paper is not with the effects of tobacco smoke on those who do the smoking. These, as indicated above, are well known and need no further discussion. Furthermore, since those who suffer the direct ill effects of smoking presumably also derive real or imagined benefits from smoking, and since they voluntarily choose to smoke, questions of wrongful risk imposition do not arise.

But recently harmful effects of smoking on nonsmokers have received considerable attention. Since these risks are by and large not voluntarily assumed and do not bring with them corresponding benefits, moral questions arise in a more interesting way. These harmful effects can be discussed in two categories: acute effects and chronic effects.

Acute Effects of Tobacco Smoke on Nonsmokers

Although the smoking habit may provide pleasure to its practitioners, many nonsmokers find that the acrid fumes of tobacco smoke cling to their hair and clothes and interfere with their enjoyment of food. In addition, many nonsmokers experience irritating effects from tobacco smoke, particularly those who have never smoked. The latter group currently comprises about 44 percent of the population (all ages). In 1979, about 80 percent of these "never smokers" (an estimated 77 million persons) reported that it was "annoying to be near a person who is smoking cigarettes." By contrast, only 5 percent of smokers found others' smoke to be objectionable.[8]

"Annoyance" from tobacco smoke takes many unpleasant forms, ranging from reaction to bad odors to serious eye, nose, and throat irritations. Nonsmokers differ in their sensitivity to tobacco smoke, with those who suffer from allergic disease (about 10 percent of the population[9]) being most likely to exhibit acute debilitating sensitivity. Speer, using a measure of nine indices of sensitivity to tobacco smoke, found that allergic nonsmokers reacted on the whole two-and-one-half times as frequently as nonallergic nonsmokers.[10] Sensitivity to tobacco smoke may produce such symptoms as itching, tearing, burning, or swelling of the eyes; sneezing, or blocking, running, itching, or dryness of the nasal airways; headache, coughing, wheezing, sore throat, hoarseness, nausea, or dizziness. Hypersensitive nonsmokers reported particular increases in the frequency of nose and throat irritation, and in wheezing, relative to nonallergic nonsmokers. Ten percent of the 44 percent of the population who never smoked might be expected to be allergic and therefore potentially hypersensitive to tobacco smoke. In fact, a Gallup Poll taken in 1978 reported that about 4 percent of nonsmoking respondents professed "allergy" to tobacco smoke.[11] Thus, an estimated 4 percent of the nonsmoking population (about 3 percent of the total population or about one out of every 33 persons) appears to be hypersensitive to ambient tobacco smoke. A recent study by Dahms et al. compared ten patients with bronchial asthma to ten normal controls under passive smoking conditions well within the range commonly reported for taverns and nightclubs; after one hour of smoke exposure, three major indices of pulmonary function had (reversibly) decreased by an average of 20 percent.[12] Dahms et al. concluded that bronchial asthmatics (about 4 percent of the population, interestingly[13]) are at risk from ambient tobacco smoke.

Savel, in a study of eight nonsmokers with acute hypersensitivity to tobacco smoke, found a sixfold enhancement of acute lymphocyte stimulation, moderate to intense upper respiratory discomfort, headaches in 50 percent of the group, with symptoms beginning typically within 30 minutes to one hour of onset of exposure and persisting for at least 8 to 12 hours.[14] Aronow found that passive smoking by ten patients with angina produced dynamic changes in blood pressure and decreased time to onset of anginal attack by 22 percent and 38 percent respectively under conditions of good and poor ventilation.[15] Thus, the toxic effects of ambient tobacco smoke on those with preexisting disease states are manifest. Its effects on healthy normals, however, have only begun to be studied.

Chronic Effects of Tobacco Smoke on Nonsmokers

It has long been known that high levels of smoke from factory chimneys could cause illness and death during air pollution episodes and that elevated levels of smoke were epidemiologically implicated in chronic respiratory morbidity. For this reason, laws such as the Clean Air Act were passed in order to protect human health from the effects of outdoor air pollution, and National Ambient Air Quality Standards (NAAQS) were established. In the establishment of such standards it was explicitly assumed that air contaminants external to buildings was the only concern, and that if this could be cleaned up, public health would be protected from the ravages of air pollution.

It fell to a physicist and a chemist at the Naval Research Laboratory in Washington, D.C., to demonstrate, in an after-hours research project, what the air pollution establishment would not or could not see: that the smoke pollution inhaled indirectly from cigarettes, pipes, and cigars indoors was not only chemically related to the smoke from factory chimneys, but routinely occurred at far higher levels indoors than did factory smoke or automobile exhaust outdoors.[16] The Naval researchers' controlled experiments and field studies showed that in buildings where tobacco is smoked, substantial air pollution burdens were inflicted upon nonsmokers, far in excess of those encountered in smoke-free indoor environments, outdoors, or in vehicles on busy commuter highways. Daily exposure to ambient tobacco smoke, they found, could cause air pollution levels corresponding to violation of the annual National Ambient Air Quality Standard for Total Suspended Particulates for exposed office workers, at typical building occupancies and ventilation rates, and amounted to the single most important source of exposure of the population to this harmful kind of air pollution.

Furthermore, based on modeled absorbed doses, there was good reason to believe that nonsmokers might be exposed to the risk of smokers' diseases from routinely breathing indoor air pollution from tobacco smoke.[17] In fact, as this heretical research was going to press, a physiologist and a physician at the University of California in San Diego published a paper in the *New England Journal of Medicine* which made coast-to-coast headlines: after ten years of research on more than 2,000 persons, they had concluded that long-term exposure to indoor tobacco smoke in the workplace was deleterious to the healthy normal nonsmoker and significantly reduced small airways function to the same extent as smoking one to ten cigarettes per day![18] The *New York Times* editorialized in a piece entitled "Thy Neighbor's Lungs," "That is not yet proof that the reduced lung function in an otherwise healthy body is medically important. Most victims were probably unaware of the impairment. But deterioration of the small airways in the lung does often precede lung disease. It is another reason, on top of earlier reasons, to prohibit smoking in indoor public places unless the smokers can be segregated so that they cannot jeopardize their neighbors."[19]

Nearly a year later, the chief epidemiologist of the National Cancer Center Research Institute in Tokyo, Dr. Takeshi Hirayama, published an electrifying study of lung cancer mortality in 142,857 women which reported that nonsmoking women with smoking husbands developed lung cancer at a rate nearly double that of women whose husbands did not smoke.[20] The findings received worldwide attention. Sir Richard Doll, an internationally prominent British epidemiologist, commented that Hirayama's study was scientifically sound; the implication, he said, is that cigarette smoking poses "a hazard to anybody in public rooms if they are not well ventilated."[21] Four days later, the *New York Times*, in a piece entitled "Smoking Your Wife to Death," would editorialize, "So much for the notion that second-hand smoke is merely a nuisance. The Japanese study, published in the authoritative *British Medical Journal*, adds to the growing evidence that second-hand smoke kills. The results strengthen the case for banning smoking in public places, especially where abstainers are exposed to smoke for long periods."[22] The Tobacco Institute initially refused comment, pleading that they had not seen the study. But behind the scenes, they retained the nationally prominent statistician Dr. Nathan Mantel to find flaws in Hirayama's methods. In a memo to the Tobacco Institute, Mantel speculated that Hirayama had possibly made an error in a standard test of statistical significance; the Tobacco Institute widely publicized this speculation, but Mantel, in an interview, denied reaching a conclusion.[23] A voluminous correspondence attacking and defending Hirayama filled the letters column of the *British Medical Journal*. The Journal went to unusual lengths to air the controversy, even publishing a letter from the Tobacco Institute.[24]

Meanwhile, two new papers added further fuel to the controversy. Appearing almost simultaneously with the Japanese work was a small and little-publicized but well-crafted study of a few hundred Greek women by Trichopoulos et al.[25] Like Hirayama, the authors also found that the nonsmoking wives of smoking husbands suffered lung cancer at approximately double the rate as their counterparts with nonsmoking husbands. However, five months later, Lawrence Garfinkel of the American Cancer Society published a paper in the *Journal of the National Cancer Institute* which examined lung cancer in 176,739 nonsmoking women as a function of their husbands' smoking habits.[26] Although a small positive result was obtained, it did not attain statistical significance, and no dose-response relationship was observed. Garfinkel concluded that any effect of passive smoking on lung cancer was very small. His analysis did not attempt to control for confounding variables, however, and the study was not originally designed to detect an effect of passive smoking.[27] Nevertheless, the Tobacco Institute widely trumpeted the study as a refutation of Hirayama's work, and the *Medical World News* carried the headline, "On Passive Smoking: Cancer Society and Tobacco Institute in Rare Unity."[28] The U.S. Surgeon General, after weighing the rival studies, concluded:

"In recent months, the popular press has generated interest in the controversy of whether passive or involuntary smoking causes lung cancer in nonsmokers. Three epidemiological studies examined this issue in the past year. . . . While the nature of this association is unresolved, it does raise the concern that involuntary smoking may pose a carcinogenic risk to the nonsmoker. . . . Therefore, for the purpose of preventive medicine, prudence dictates that nonsmokers avoid exposure to second-hand tobacco smoke to the extent possible."[29] Dr. Nicholas J. Wald, a noted Oxford University cancer epidemiologist, went even further, stating that reports that the Cancer Society study contradicted the other two were "absolute rubbish." He warned that "BOPS—breathing other people's smoke—may bring early death to as many as 2,000 nonsmoking Americans every year."[30]

Thus, the issue is drawn: there is now evidence, some of it disputed, that ambient tobacco smoke, like ambient factory emissions, increases the risk of cancer and respiratory disease.

Obstacles to Risk-Aversive Action

In the next section I consider different strategies for regulating and controlling the risks posed by passive cigarette smoking. But several factors conspire to make resolution of this problem difficult.

First, those suffering the acute debilitating effects of passive smoking, as the statistics cited above indicate, are a small minority. Since hypersensitives make up only an estimated 3 percent of the general population, this means that in many social and professional situations the hypersensitive nonsmoker is vastly outnumbered by both smokers and moderately sensitive nonsmokers.

This group of hypersensitives is pitted, furthermore, against a powerful adversary in the form of the cigarette industry. The cigarette industry is a relatively small but highly profitable industry of six companies, 39,000 employees, and 14 manufacturing plants. Total consumer expenditures in 1981 were $15.8 billion, of which $6 billion went for taxes. The advertising budget of the industry was about $800 million. The penetration of the industry into the print advertising market accounted for 10 percent of magazine advertising dollars, 14 percent of product advertising in newspapers, 31 percent of all outdoor advertising, and 52 percent of all dollars spent in newspaper supplements. With this massive advertising budget comes influence: replacement of the loss of this large revenue would be difficult for any of these media.[31] In fact, with the exception of a few publications (e.g., *Good Housekeeping, Reader's Digest, The New Yorker*), mass-market magazines generally run extensive advertising for cigarettes, and in the case of major women's magazines, a study has indicated that adequate editorial coverage is not provided on the risks of cigarette smoking; i.e., the economic impact of advertising could be dictating the editorial policies of some magazines.[32] Moreover, public service information on smoking provided by the TV networks is "virtually invisible."[33]

The tobacco industry is well served by its ardent lobby, the Tobacco Institute, an industry trade association which serves as the propaganda arm of tobacco interests. The stated goal of the Tobacco Institute is

> to preserve the ability of business to enter into the free marketplace, . . . to create a climate in which our member companies can compete without unwarranted restraints. This means that we assist the nation's news media, its public policy setters, and the public itself in separating fact from fiction concerning smoking and health. It means pointing out the gaps in scientific knowledge as well as . . . overstatements of what is known. It means emphasizing the danger of accepting fallacy for fact in any scientific dispute before all information is available.[34]

These goals are furthered by the Institute's three professional speakers, its movie shorts, and its publications, which include its house organ, *The Tobacco Observer*. Through these, the Tobacco Institute has launched a major propaganda campaign, using a "freedom of choice" theme, to cast nonsmokers' rights advocates as "fumaphobes" who have an extreme and irrational fear of tobacco smoke. They hope in this way to persuade the silent majority of nonsmokers to dissociate themselves from the movement.[35]

As an indication of the mendacious nature of this campaign, consider the following principles gleaned from a recent issue of *The Tobacco Observer*.[36] With respect to smoking:

> Business has an unrestrained right to market products.
> Smokers have a right to personal choice in the matter of social custom, free of regulation.
> The relationship between smoking and disease cannot be proved until all information is available.

(Business, of course, does not have the unrestrained right to market harmful products; regulation of tobacco products under every conceivable federal law is specifically proscribed, however, due to the efforts of the tobacco lobby.[37] Although smokers have enjoyed the right to consume tobacco, other dangerous drugs are strictly regulated by society. Thousands of research reports now incriminate tobacco smoke as a cause of disease.) With respect to smoking restrictions:

> Antismoking groups are prohibitionists who wish to criminalize the act of smoking in public.

Smoking restrictions are costly to business and unenforceable.
Freedom is the right to choose, including the right to make choices others disapprove.

(In fact, few nonsmokers' rights advocates propose prohibition of tobacco products; most seek only restrictions on smoking. Smoking restrictions in fact work extremely well wherever they are tried, e.g., on transport aircraft, where 70 percent of passengers choose nonsmoking sections. Freedom to make choices of which others disapprove does not extend to harming their health.) With respect to environmental tobacco smoke:

Present studies of the effects of ambient tobacco smoke are inconclusive; more research is needed.
Tobacco smoke is not a major source of workplace pollution; dust and "exhaust" are much greater.
There is no proof that smoking in public affects persons with respiratory disease.
Tobacco smoke has been shown to be a minor annoyance at most.
Smoking restrictions cannot be justified on the basis of minor annoyance.
The whole issue can be dealt with by mutual courtesy.

(The failure of courtesy, inherent in the addictive nature of the smoking habit, is the source of the problem; most air pollutants are regulated on the basis of far weaker evidence than exists against ambient tobacco smoke; that smoking in public does not affect persons with respiratory disease is contradicted by many studies.[38])

Thus nonsmokers have to contend not only with opposition on a personal level from irate smokers who resent interference with their addiction, but with institutional opposition from the industry. As we shall see in what follows, this opposition has been instrumental in limiting the relief available to nonsmokers whose rights have been violated. Nonsmokers have been fighting back, however. Around 1970, two antismoking organizations were formed by nonsmokers who suffered acute effects from tobacco smoke: ASH, Action on Smoking and Health, and GASP, Group Against Smokers' Pollution. ASH, a legal action group, has been successful in bringing actions which resulted in "equal time" for public-interest and antismoking commercials on television, and in persuading the Civil Aeronautics Board to establish nonsmoking sections on aircraft. GASP, more of a grass-roots citizens' organization with chapters in many states, has been effective in lobbying for smoking-restrictive legislation. Both ASH and GASP exert pressure on governments and private businesses to restrict smoking, bring lawsuits on behalf of individual nonsmokers, educate the public on nonsmokers' rights, publish lists of business establishments restricting smoking, and distribute buttons, business cards, signs, and bumper stickers with antismoking slogans.[39] The major voluntary health organizations in the past have been criticized for their inaction on nonsmokers' rights, but recently the American Lung Association has taken an aggressive stance. These passive smoking risk-evaluating organizations, however, do not have the financial resources of the Tobacco Institute. The federal government maintains an Office on Smoking and Health, but severe budget cuts have limited its ability to counter the propaganda of the Tobacco Institute.[40] Nevertheless, there is evidence that the tobacco industry has much to fear from nonsmokers' rights groups: A Roper Poll taken for the industry in 1978 reported that the finding that nearly 60 percent of the population believes that smoking is dangerous to nonsmokers' health is "the most dangerous development affecting the viability of the tobacco industry that has yet occurred."[41]

Strategies for Approaching the Problem

In this section I survey different approaches to resolve the problem of the risks posed to nonsmokers by side-stream cigarette smoke. It is interesting that the control methods most widely suggested have not been aimed at the source of the problem—the cigarette—but at strategies for modifying relevant human behavior: through banning or restricting smoking in public places or directing nonsmokers to avoid smoky places. I consider in turn a) regulation of cigarettes; b) smoking restriction laws; c) remedies provided by the legal system; and d) attempts to regulate cigarette smoke as air pollution.

Regulation of Cigarettes

An attractive option to controlling side-stream smoke from cigarettes would be to require the marketing of cigarettes which had lower emissions, for example a self-extinguishing cigarette. However, regulation of tobacco products is specifically proscribed under the Federal Hazardous Substances Act, the Consumer Product Safety Act, the Fair Packaging and Labeling Act, the Controlled Substance Act, and the Toxic Substance Control Act. Under a recent federal court ruling, tobacco products may not be regulated under the Food, Drug, and Cosmetics Act—even though nontobacco cigarettes may be. Because federal law is preemptive, states may not regulate tobacco products. Although the Congress has imposed a federal tax on cigarettes, currently 16 cents per pack, it returns some of this revenue in the form of tobacco-growing subsidies to farmers. As a result of the Surgeon General's Report on Smoking and Health of 1964, Congress has required a health warning to be placed on all cigarette packages sold in the United States and on all cigarette advertising, and the tobacco industry voluntarily withdrew TV advertising after it lost a legal battle which required TV outlets to air "equal-time" antismoking messages. Thus, although some small progress has been made, regulation of cigarettes appears to be very difficult, a testimonial to the power of the tobacco lobby.

Legislation Restricting Smoking

A second approach is to pass legislation restricting smoking in specified areas. Swingle has reviewed such legislation: as of 1980, 34 states have legislation which restricts smoking in some fashion, and one state has accomplished this administratively.[42] All of these statutes have been enacted only since 1974 and affect one or more of the following locations: restaurants, retail stores, public conveyances, educational facilities, hospitals, nursing homes, auditoriums, arenas, offices, and meeting rooms. Twenty-five states can be described as having reasonably comprehensive laws. Seven states and the District of Columbia have very weak laws, affecting typically only such places as buses and elevators. Thirteen states have laws prohibiting smoking in schools and other areas because of fire hazards or food contamination. Three states—Alabama, North Carolina, and Wisconsin—have no laws restricting smoking. Among the stronger laws are those which declare smoking to be a health hazard or public nuisance; a few even state that their intent is to protect the right of the nonsmoker to breathe clean air. Only four states—Minnesota, Montana, Nebraska, and Utah—have statutes regulating smoking in the workplace. The Minnesota statute is representative; it provides that it is the employer's responsibility to provide a smoke-free area at the employee's request and even holds out the possibility of injunctive relief for victims of repeated violations; violation of the statute is ordinarily treated as a petty misdemeanor. In most states, however, workers suffering from smoking on the job have had only the unattractive alternatives of continuing to suffer, quitting their jobs, taking a disability retirement, or taking legal action.

Strong statewide laws such as Minnesota's are extraordinarily difficult to pass as evidenced by their rarity and by the following record of failed attempts. In 1979, 38 states introduced 116 bills to limit or extend restrictions against smoking in various public places. Seven were passed by seven states. In 1980, only one out of 49 such bills passed, generally due to the determined opposition of the tobacco lobby.[43] The author has been present at legislative hearings in eight jurisdictions in the metropolitan

Washington, D.C., area; in each instance, one or more representatives of the tobacco industry were present and argued against the bill (nonsmokers' rights advocates in several states confirm that this is a usual phenomenon). In four of the eight cases (where there was a large public turnout) most of the legislation was passed, but often only after several tries. However, it is comparatively much easier to pass legislation on a county level, as evidenced by the stunning series of 30 successes achieved in California recently.

Remedies Through the Legal System

Secondhand smoke in the workplace has provoked a number of important lawsuits through which afflicted nonsmoking workers have sought relief from an intolerable work situation. The first three suits involve federal employees, who generally have different legal rights from private business employees.

(1) Labor-management strife at the Social Security Administration in Baltimore led to a study by Barad of 10,000 nonsmoking federal workers.[44] Barad found that ambient tobacco smoke impaired the working efficiency of more than 50 percent of the nonsmokers to some extent, with 36 percent reporting that they were forced to move away from their work stations to avoid breathing smoke. Twenty percent of the nonsmokers reported difficulty in concentrating on work due to passive smoking, and 14 percent found it difficult to produce work as a result of smoke on the job. Nearly 25 percent of the nonsmokers expressed feelings of frustration and hostility toward smokers and management. Ten out of these 10,000 employees ultimately formed an organization called the "Social Security Administration Ecology Group Committee on Tobacco Smoke and Its Hazardous Effects on Employee Health." Allied with two nonprofit nonsmokers' rights organizations (GASP and ASH), 17 other federal workers, and a group called FENSR (Federal Employees for Nonsmokers' Rights), they brought suit against the federal government in the District of Columbia Circuit Court on June 17, 1977, because of inability to obtain relief through normal administrative channels. It is instructive to inquire into the details.

According to Barad, a full 10 percent of the employees reported "allergy" to tobacco smoke, and an additional 3 percent reported aggravation of a cardiovascular disorder. If just the allergics are considered, this amounts to about 4.5 percent of the work force, in fair agreement with our earlier estimate of about 3 percent of the general population being hypersensitive to tobacco smoke.[45] In the case of the Social Security Administration, the hypersensitive group was outnumbered by about seven to one by the smokers (31 percent of employees) and by 15 to one by both smokers and nonsmokers who did not have any problem with ambient tobacco smoke. Moreover, Barad reports that 41 percent of the supervisors smoked, further complicating the problem for the hypersensitives in the rank-and-file. In fact, 52 percent of those nonsmokers who tried informally to get smoking coworkers to refrain from smoking in the office were generally unsuccessful, and another 34 percent got mixed results. When polled, 55 percent of the smokers rejected special designated smoking areas away from the workstation, rejected separation of smokers and nonsmokers into different work areas, and rejected a total prohibition of smoking at work (the latter was favored by 33 percent of the nonsmokers).

Having exhausted their administrative remedies, the bitterly frustrated core of hypersensitives went to court, asserting that the federal government had a common-law duty to protect the health of its employees. Nearly two years later, U.S. District Court Judge Charles R. Richey dismissed the case with prejudice, stating that health protection of employees was not appropriately a part of the common law, and that the Court therefore lacked jurisdiction.[46] This dismissal was upheld by the Court of Appeals and ultimately by the Supreme Court itself in its October 1979 term. Thus, these public employees were told in effect that their right to enjoy a smoke-free workplace was not legally guaranteed (Figure 1.2).

(2) This principle was enunciated more clearly in September 1982, when Trial Judge Donald Voorhees ruled that Lanny Vickers, a purchasing agent for the Veterans' Administration Hospital in Seattle, who developed headaches and severe eye, nose, and throat symptoms from exposure to ambient tobacco smoke, was a handicapped person under the Federal Rehabilitation Act of 1973. The judge further ruled, however, that the Veterans' Administration could not be forced to provide him with a smoke-free working environment. According to Judge Voorhees, who said he had never smoked, smokers in a workplace "have certain rights" which must be balanced against Vickers's desires. "Until and unless Congress enacts a statute banning the smoking of tobacco in Government offices or the Veterans' Administration promulgates a policy against smoking in its offices," the judge asserted, "the desires of those employees who wish to smoke cannot be disregarded." The judge said that it was Vickers's responsibility to avoid tobacco smoke. The Labor Department, however, has awarded Vickers compensation for the time he misses from work because of the ill effects of tobacco smoke.[47]

(3) On the heels of the Vickers case, in October 1982, the 9th U.S. Circuit Court of Appeals in San Francisco extended Judge Voorhees's doctrine in a direction more favorable to nonsmokers. Irene Parodi, a procurement clerk with the Defense Logistics Agency, became ill several years earlier after being moved to an office with many heavy smokers.

After unsuccessfully trying to get a transfer away, she appealed for a disability retirement on the grounds that she could not work around smokers. The Office of Personnel Management turned her down on the grounds that her condition did not meet the government's definition of disability; the decision was reaffirmed by the Merit Systems Promotion Board. Parodi then took her case to court. The Appeals Court ruled that the Defense Department either must find and offer Parodi "suitable employment in a safe environment," or it would order the government to retire her on disability.[48] On August 1, 1984, a stipulated settlement was reached where Parodi was awarded a lump-sum payment of $50,000 and a monthly disability payment, with the judge ruling that the government had failed to provide a safe locale. This decision, while helpful to hypersensitive nonsmokers on the federal payroll, falls far short of providing sufficient protection to nonsmokers, since a finding of disability does not require accommodation, while a finding of "handicapped" would have required the government to provide relief on the job.

What of the situation for private sector employees? It might be expected that bringing a legal action against a private employer would be considered legitimate grounds for dismissal—a problem not faced by the federal workers. Let us consider three cases brought against private employers which illustrate the private workplace problem.

(1) Donna Shimp, an employee of New Jersey Bell Telephone Company for nine years, was transferred to a small unventilated office containing seven smokers out of 13 workers where, in her own words, she felt like "a piece of meat thrust into a smoke-house for curing." She developed erosion of the cornea, skin rashes, vomiting, and headaches. On her physician's advice, she remained home for three months on disability. Upon her return to work, a company physician advised that she return only to a smoke-free environment. A company executive told her that such an environment could not be provided in the department where she worked, and that neither the company nor the union would consider a change in policy. She was told that her alternatives were to continue to work in an environment that was damaging her health, to be transferred with a demotion and salary cut to a nonsmoking area designed to protect sensitive switching equipment, or to be fired to protect her health. She appealed this decision and exhausted her administrative remedies with both the company and the union, seeking to restrict smoking or to provide increased ventilation. She then appealed for assistance to the Environmental Protection Agency, the Occupational Safety and Health Administration, and the National Labor Relations Board, with a similar lack of success. Frustrated, she turned to the courts. A law school professor agreed to represent her without fee. Two years later, in 1977, the New Jersey Superior Court enjoined New Jersey Bell to restrict Shimp's

Figure 1.2
Reprinted with special permission of King Features Syndicates, Inc.

coworkers' smoking to a nonwork area.[49] The suit was based on the legal theory that it is an employer's duty to provide safe working conditions. The case never went beyond trial court, however, and is therefore considered less than definitive.[50] Plaintiffs in two more-recent cases have not fared as well.

(2) Paul Smith, an employee of Western Electric in St. Louis, had tried in vain for five years to convince his superiors that working in a smoke-filled room with 30 smokers was responsible for his headaches, high blood pressure, nausea, chest pains, and a litany of other medical complaints.[51] Finally, three weeks of tests in a specialized clinic produced medical evidence of his intolerance to ambient tobacco smoke. The company, convinced of his medical problem, gave Smith the enforced alternatives of either wearing a respirator on the job (so that the company could avoid liability), or accepting a transfer to a smoke-free computer area — at a $500 per month pay cut. In October 1980, after wearing the bulky, noisy respirator for several months, Smith filed suit against the company, demanding that smoking be barred in all but designated areas of his office.[52] Circuit Court Judge Phillip J. Sweeney threw out the case, asserting that Smith did not have a remedy under Missouri law. He appealed this decision, and in September 1982, the Missouri Appellate Court, in a precedent-setting opinion, remanded the case back to the trial court, asserting that Missouri common law does require an employer to provide a reasonably safe workplace, and that the plaintiff is not only entitled to a trial on the merits, but if his case is proved, is entitled to relief.[53] Win or lose, however, this case will cost Smith several thousand dollars in legal fees. At this writing, Smith is, on advice of his physician, on an extended sick leave and in August 1984 completed his second trial.

(3) Adele Gordan, an employee of Raven Systems and Research, Inc., a contract research firm in Washington, D.C., arranged with her supervisor to be seated in a smoke-free area upon joining the firm. After a time, her supervisor ordered her to move into an area contaminated by smoke. Gordan refused, explaining that upon exposure to smoke, her eyes watered, she coughed, and her nose became stuffy, red, and swollen. Her boss allegedly replied "the work comes before your health." She persisted, was fired, and filed suit in D.C. Superior Court.[54] After several days of hearings before a jury, the trial judge threw out the case, asserting that the plaintiff had no remedy under D.C. law. The D.C. Court of Appeals ruled in May 1983 to uphold the lower court ruling, stating that companies in the District of Columbia do not have to provide special accommodations for nonsmoking employees who wish a smoke-free working environment, and that it is up to the D.C. City Council to make laws protecting smokers. (The Council has considered such a law, but it failed to pass).[55] Gordan, like Shimp, has fortunately received the benefit of pro bono legal assistance. Unlike Shimp, she has lost her job.

Thus, with the single exception of the Shimp case, trial courts have appeared quite reluctant to impose a judicial solution to the problem of the acute effects of indoor air pollution from tobacco smoke, preferring to leave the problem to the legislative arena. This may be due to the relatively small percentage of the population exhibiting hypersensitivity to tobacco smoke, the widespread social acceptance of the custom of smoking, and the almost total lack of federal regulation of the product. However, the recent precedent-setting appellate decision in *Smith* v. *Western Electric* may encourage future nonsmokers' litigation, especially if Smith is able to prove his case.

Tobacco Smoke as Hazardous Air Pollution

Section 112 of the Clean Air Act calls for regulation of pollutants which may "reasonably be anticipated to result in an increase in mortality or an increase in incapacitating reversible illness." Airborne carcinogens are given a high priority for listing as hazardous air pollutants. Nowhere in the act is it specified that the pollutant must be an outdoor one; this is simply an historical interpretation of its mandate by the Environmental Protection Agency (EPA).

The provisional policies developed by the EPA for identifying, assessing, and regulating airborne substances which may pose a risk of cancer require two basic questions to be answered: What is the probability that the substance is a human carcinogen? What is the extent of human exposure?[56] In response to the first question, the U.S. Surgeon General has said: "Cigarette smoking is the major single cause of cancers of the lung, larynx, oral cavity, and esophagus, and is a contributory factor for the development of cancers of the bladder, pancreas, and kidney. The term contributory factor by no means excludes the possibility of a causal role for cancers of these sites."[57] As far as the extent of nonsmokers' exposure, as we have discussed, the pandemic nature of smoking makes it quite difficult for a nonsmoker to avoid exposure, from childhood onward. In 1970, a national survey indicated that nearly two-thirds of U.S. children resided in homes where at least one person smoked.[58] For adults, smoking does not appear to be prohibited in more than about 11 percent of white-collar and 28 percent of blue-collar U.S. workplaces.[59] Tobacco smoke persists in indoor spaces long after an individual cigarette is extinguished: at a typical ventilation rate of one air change per hour, it takes about three hours for 95 percent of the smoke to dissipate; at the average smoking rate of two cigarettes per hour per smoker, there is always residual smoke in a room. Since the average person spends more than 90 percent of the day indoors, and since one out of three adults smokes, many nonsmokers spend significant fractions of the day in an atmosphere contaminated by tobacco smoke.[60] Thus, ambient tobacco smoke might be considered a prime candidate for listing as a hazardous air pollutant under the Clean Air Act, based solely on the criteria of carcinogenicity and exposure.

Once a hazardous air pollutant has been identified as an airborne carcinogen, the regulatory process calls for the listed pollutant to undergo quantitative risk assessment. While cancer risk estimation is an imprecise endeavor involving many uncertainties, such estimation can often provide a rough measure of the magnitude of the carcinogenic risk of the substance.[61]

A rough estimate of the order of magnitude of the effect of passive smoking can be obtained from Hirayama's 1981 study.[62] Hirayama found that the average effect of passive smoking on nonsmoking Japanese women whose husbands smoked was to increase their lung cancer death rate by about eight cases per 100,000 population per year in the population age group at risk of lung cancer (≥ 40 years). Repace and Lowrey have estimated a similar rate in the U.S. population at risk.[63] How does such a risk compare to other risks faced by the U.S. population? Lichtenstein et al. have given a table of statistical estimates of the frequency of death per 100 million persons in the U.S. as an average over the years 1968 to 1973, shown in Table 1.1.[64] Added for comparison is the eight per 100,000 risk scaled to the U.S. population in 1970. The estimated annual risk is 3,700 lung cancer deaths from passive smoking per 10^8 U.S. residents. (This average is taken over *all* age groups, not just those at risk of lung cancer, to be compatible with the numbers presented in the rest of the table.) 1978 data is added for smokers taken as a group. It can be seen that this estimated risk is considerably in excess of many other involuntary risks and appears to be far from trivial.

Is the level of risk from passive smoking acceptable? Some risks are considered "acceptable" by society because they are purely personal and voluntary, and by some individuals because they derive some benefit, real or imagined. Examples in this category might include the risk of premature death from heart disease by being grossly overweight, or from accidental falls during mountain climbing. Some risks are considered acceptable because they are beyond the ability of society or individuals to control, such as earthquakes. Other risks are considered unacceptable by society because the risk to society is acute. An example would be experiments involving implantation of human enteral bacteria with the genes for the production of botulinus toxin. In between fall a large number of risks whose magnitude is reduced by risk-aversive action: society maintains fire and police forces; individuals wear seat belts in autos and avoid air transportation or military service. If society or individuals "consent" to a risk, they have judged it acceptable. If they do not consent to a risk, they have judged it unacceptable and may, as far as possible, take risk-

Table 1.1 Comparison of Estimated Lung Cancer Mortality from Passive Smoking with Statistical Estimates of the Frequency of Death per 10^8 Residents per Year, 1968–1973 Average

Cause	Rate/10^8
Smallpox	0
Poisoning by vitamins	0.5
Botulism	1
Measles	2.4
Fireworks	3
Smallpox vaccination	4
Whooping cough	7.2
Polio	8.3
Venomous bite or sting	23.5
Tornado	44
Lightning	52
Nonvenomous animal	63
Flood	100
Commercial air crashes	100
Excess cold	163
Syphilis	200
Pregnancy, childbirth, and abortion	220
Infectious hepatitis	330
Appendicitis	440
Electrocution	550
Motor vehicle-train collision	740
Asthma	920
Firearm accident	1,100
Poisoning by solid or liquid	1,250
Tuberculosis	1,800
Fire and flames	3,600
Drowning	3,600
Passive smoking (average)[a]	3,700
Leukemia	7,100
Accidental falls	8,500
Homicide	9,200
Emphysema	10,600
Suicide	12,000
Breast cancer	15,200
Diabetes	19,000
Motor vehicle (car, truck, or bus) accidents	27,000
Lung cancer	37,000
Cancer of the digestive system	46,600
All accidents	55,000
Stroke	102,000
All cancer	160,000
Active smoking[a,b]	350,000
Heart disease	360,000
All disease	849,000

[a]Estimated, averaged over the entire 1970 U.S. population (including those not at risk; the risk is 8000 per 10^8 in the nonsmoking population of cancer age, and 650,000 per 10^8 in adult smokers).
[b]Based on 1978 figures.
Source: Adapted from Lichtenstein et al.

aversive action until the perceived "cost-benefit" ratio is acceptable. This usually means that risks are not entirely eliminated, i.e., reduced to zero, but rather reduced to the limit to which society or the individual is willing to pay.

Table 1.2, adapted from the work of Dinman and Rowe, gives the crude "boundaries of acceptable risk."[65] Fatal risks greater than 10^{-2} per year are generally unacceptable to society, and risks below 10^{-11} are below the threshold of concern. As a practical matter, risks below 10^{-6} per year are often considered acceptable by most individuals and tend to be below the

Table 1.2 Boundaries of Acceptable Risk

Annual Per Capita Risk	Societal Response, Degree of Acceptability
10^{-2}	Fatal illness and disease rates: society maintains expensive preventive and therapeutic medical institutions to mitigate risk. Smoking risk.
10^{-3}	Fatal accident rates: infrequent at this level; immediate action taken to reduce hazards.
10^{-4} Passive smoking (estimated average)	Fatal accident rates: public money is spent to control causes, police and fire depts. are supported, traffic controls are imposed. Average level of regulated voluntary risk by OSHA.
10^{-5}	Threshold of regulated voluntary risk. Society warns against dangers of fire, drowning, poisons.
10^{-6}	Threshold of regulation for hazardous air pollutants. Fatal accidents at this level are not of concern to most individuals (level of risk of death by earthquake in California, and of death in a civilian air crash). Society prescribes building codes for earthquake resistance, regulates commercial air travel. Proposed Nuclear Regulatory Commission Safety Goals for Nuclear Plant Accidents.
10^{-7}	Level of natural catastrophic involuntary risk — tornado, hurricane, lightning; weather forecasting services maintained, lightning protection prescribed in building codes.
10^{-8} 10^{-9}	Food and Drug Administration Guideline "upper bound"; regulation of ordinary involuntary risk from oil, gas, coal, and nuclear power industries.
10^{-10}	Regulation of catastrophic risk from energy-producing industries.
10^{-11}	Threshold of concern (loss of life expectancies of the order of 10^{-5} days).

societal threshold of risk aversion, with the exception of certain catastrophic risks that have generated public fear. In the region between 10^{-2} and 10^{-6}, society may be willing to accept certain levels of risk higher than 10^{-6} because further reduction is beyond its ability to pay. In essence, a risk-benefit trade-off is made. This is typical for the risks associated with certain industrial activities, which, although they emit air pollutants, are perceived to be essential in some degree to society. In cases where risks cannot be quantified in this manner, federal regulatory agencies will often try to reduce risks by requiring "best available control technology" or "reasonably achievable control technology," defined by the perceived nature of the risk and ability of the regulated industry or society to pay. Quantitative risk assessment, where possible, is used to convince both the regulated industry and the public that the regulatory agency is not being arbitrary.

For example, the U.S. Food and Drug Administration has regarded risks of the order of 10^{-8} per year for carcinogenic residues in foods as "virtually safe." The Environmental Protection Agency has regarded cancer risks of the order of 10^{-7} to 10^{-6} per year as an approximate guideline for the regulation of carcinogenic hazardous air pollutants. The Nuclear Regulatory Commission has proposed guidelines for acceptable cancer mortality risks from reactor accidents of the order of 10^{-6} per year. The Royal Commission on Environmental Pollution in the U.K. has suggested that individual risks below 10^{-6} per year are "acceptable," whereas risks above 10^{-2} per year are "unacceptable." Judged by these guidelines of societal acceptability for carcinogenic risks, an involuntary lung cancer risk of nearly 10^{-4} per year from passive smoking would appear to require regulation.

On these grounds, then, society should regulate the risks of passive smoking. First, the risks of passive smoking are apparent; second, the benefits of passive smoking are not apparent; finally, a solution is available at virtually no cost: separate sections for smokers and nonsmokers in public places. This solution might be considered "reasonably achievable control technology"; although it does not eliminate the risk as would a complete ban, it can be achieved at relatively little social cost. Figure 1.3 illustrates the problem, and Figure 1.4 one solution.

In the workplace, a complete physical separation is required because of the cancer risk and to accommodate hypersensitives, but "best available control technology" should be considered here: complete bans on smoking on the premises. As a number of studies have shown, this solution benefits not only nonsmokers; it benefits smokers and results in dollar savings to employers as well, because the higher sick-leave and housekeeping costs caused by smokers disappear; an increasing number of industries are trying this approach.[66]

Are such measures practical? A survey of 3,000 U.S. corporations in

Figure 1.3
From *Herblock on all Fronts* (New American Library, 1980).

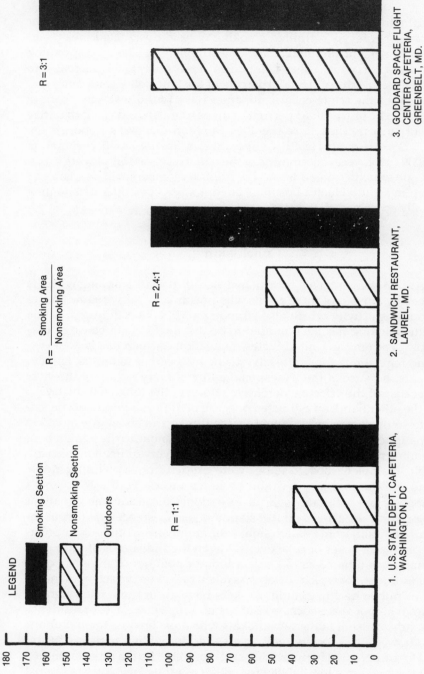

Figure 1.4
Relative Indoor/Outdoor Air Pollution Levels

1979 (29 percent response rate) indicated that 42 percent of blue-collar companies surveyed permitted smoking only in designated areas, and another 28 percent prohibited smoking completely. The corresponding percentages for white-collar companies were poorer, 15 percent and 11 percent respectively. Thus 70 percent of the blue-collar companies and 26 percent of the white-collar companies have found restrictive control measures to be practical; 65 percent of the respondees indicated that they had had such policies since the 1964 Surgeon General's Report.[67] On April 1, 1984, Group Health Cooperative, a Seattle-based corporation with 5000 employees, became one of the nation's largest employers to ban on-the-job smoking; in addition, the 325,000 members will be unable to light up in Group Health hospitals or clinics, except under extenuating medically approved circumstances.[68]

Conclusion

Cigarette smoking is a well-established vice of 70 years standing which is practiced by one out of three adults who presumably derive some "benefit" from this activity which also supports a $15 billion industry. While cigarette smoking has been judged to be the single major cause of preventable death in the United States, regulation of cigarettes is nonexistent, and regulation of cigarette advertising has been de minimus, largely, it appears, because of the widespread and voluntary nature of the vice, and because of the effective defensive efforts of the tobacco industry. On the other hand, an estimated one-half of the adult population is irritated to some degree by ambient tobacco smoke and about one out of 33 persons appears to be hypersensitive to tobacco smoke to the point where functioning normally in a smoky environment is very difficult or even impossible. If ambient tobacco smoke were emitted from a polluting industry into the outdoor air, it would be judged to be both a toxic and a carcinogenic air pollutant, subject to national hazardous air pollutant emission controls. Thus a double standard is in existence that judges indoor air pollution from tobacco smoke differently from outdoor air pollution from diesel buses or coke ovens. Given the similarities of indoor air pollution from tobacco smoke to outdoor air pollution from other combustion sources, for which society has taken risk-aversive action in the interest of public health, should smokers have an untrammeled right to pollute an indoor space such as a public building, workplace, or residence without the consent of the nonsmokers who must breathe the pollution? If yes, how should society, employers, and individual nonsmokers respond to the potential for morbidity and mortality? As the asbestos industry is facing massive tort liability for failing to protect its workers against the hazards of asbestos, similarly the recently emerged scientific knowledge of the health risks and air pollution levels generated by ambient tobacco smoke may establish a nexus between passive smoking health risk to workers and economic risk to employers. As we have seen, nonsmokers are beginning to assert tort claims against employers. At the least, if society yields smokers untrammeled rights to smoke in public indoor spaces, including workplaces, should it not require increased ventilation or air cleaning measures in such buildings, using the same logic used to limit criteria air pollutants from factory smokestacks and automobiles? Should it not require self-extinguishing cigarettes, control of cigarette additives, and limitation on side-stream tar and nicotine content, using the same logic used to control carcinogenic air pollutants from chemical plants? Isn't it time to treat ambient tobacco smoke as the hazardous air pollutant it is, and isn't it time to subject the tobacco industry to the same sort of controls that *all* other polluting industry must bear?

Finally, does the individual smoker have the moral right to cause harm in the form of physical irritation or carcinogenic risk to the nonsmoker? If so, how should smokers accept the following proposition: Suppose that individual nonsmokers, in defense of their asserted right to breathe tobacco-smoke-free indoor air, were to release a gas into indoor spaces where they were forced to breathe tobacco smoke. Suppose further that when sucked through the burning cone of a cigarette, pipe, or cigar this gas decomposed into irritating byproducts which caused moderate to intense discomfort to the smoker, much the way ambient tobacco smoke affects the nonsmoker. Would smokers feel that they had the inalienable right to gas-free air?

I assert that smokers have the right to enjoy the risks and benefits of smoking, just as they have the right to play Russian roulette. When they smoke indoors in the presence of nonsmokers, however, they are playing Russian roulette with nonsmokers' health. This they do not have the right to do. Therefore either smokers should abstain from smoking indoors or obtain explicit consent from each exposed nonsmoker (or from the parents or guardians of minors and infants). Smokers can refrain from smoking, smoke outside, or quit. Nonsmokers cannot refrain from breathing. Smokers' pollution harms nonsmokers; nonsmokers' breathing does not harm smokers. Smokers are on the same moral ground as were spitters around the turn of the century when public health laws restricted spitting in public buildings. The nonsmoker's pursuit of healthful breathing air does not infringe upon the health of smokers (in fact it may improve it to the extent that it forces them to cut down or to quit the habit). Thus, nonsmokers' right to clean indoor air is morally superior to smokers' right to pollute. As the awareness of passive smoking risks grows in nonsmokers, smokers will be increasingly, and accurately, viewed as playing a hardnosed game with nonsmokers' health, a practice which can only invite increasing confrontation and social dissention.

Notes

1. Discussions of the social history of tobacco smoking can be found in F. Speer, "Tobacco and the Nonsmoker: A Study of Subjective Symptoms," *Archives of Environmental Health* 16 (1968); U.S. Surgeon General, *The Health Consequences of Smoking for Women—A Report of the Surgeon General* (Washington, D.C.: U.S. Dept. of Health and Human Services, 1981); U.S. Surgeon General, *Smoking and Health—A Report of the Surgeon General* (Washington, D.C.: U.S. Dept. of Health, Education, and Welfare, 1979); E. Eckholm, "Cutting Tobacco's Toll," Worldwatch Paper 18 (Washington, D.C.: Worldwatch Institute, 1978).
2. See J. D. Trop, *Please Don't Smoke in Our House* (Chicago: Natural Hygiene Press, 1976).
3. U.S. Surgeon General, *Smoking and Health*.
4. J. L. Repace and A. H. Lowrey, "Indoor Air Pollution, Tobacco Smoke, and Public Health," *Science* 208 (1980); U.S. Surgeon General, *Smoking and Health*.
5. U.S. Surgeon General, *Smoking and Health*.
6. U.S. Surgeon General, *The Health Consequences of Smoking* (Washington, D.C.: U.S. Dept. of Health, Education, and Welfare, 1964).
7. U.S. Surgeon General, *Smoking and Health*.
8. Ibid.
9. NIAID Task Force Report, *Asthma and Other Allergic Diseases* (Washington, D.C.: U.S. Dept. of Health, Education, and Welfare, 1979).
10. Speer, "Tobacco and the Nonsmoker."
11. The Roper Organization, Inc., "A Study of Public Attitudes Toward Cigarette Smoking and the Tobacco Industry," Vol. 1, May 1978, Secret public opinion poll taken by Roper for the tobacco industry disclosed by the Federal Trade Commission.
12. T. Dahms, J. Bolin, and R. Slavin, "Passive Smoking: Effects on Bronchial Asthma," *Chest* 80 (1981).
13. G. Greene, "Nonsmokers' Rights: A Public Health Issue," *Journal of the American Medical Association* 239 (1978).
14. H. Savel, "Clinical Hypersensitivity to Cigarette Smoke," *Archives of Environmental Health* 21 (1970).
15. W. Aronow, "Effects of Passive Smoking on Angina Pectoris," *New England Journal of Medicine* 229 (1978).
16. Repace and Lowrey, "Indoor Air Pollution."
17. Ibid.
18. J. R. White and H. F. Froeb, "Small Airways Dysfunction in Nonsmokers Chronically Exposed to Tobacco Smoke," *New England Journal of Medicine* 302 (1980).
19. Editorial, "Thy Neighbor's Lungs," *New York Times*, April 2, 1980.
20. T. Hirayama, "Nonsmoking Wives of Heavy Smokers Have a Higher Risk of Lung Cancer: A Study from Japan," *British Medical Journal* 282 (1981).
21. L. K. Altman, "Cancer Study Reports Higher Risk for Wives of Smoking Husbands," *New York Times*, January 16, 1981, p. A1.
22. Editorial, "Smoking Your Wife to Death," *New York Times*, January 21, 1981.
23. "On Passive Smoking: Cancer Society and Tobacco Institute in Rare Unity," *Medical World News*, July 6, 1981, p. 30.
24. H. Kornegay and M. Kastenbaum, "Letter to British Medical Journal," *British Medical Journal* 283 (1981): 914.
25. D. Trichopoulos, A. Kalandidi, L. Sparros, and B. MacMahon, "Lung Cancer and Passive Smoking," *International Journal of Cancer* 27 (1981).
26. L. Garfinkel, "Time Trends in Lung Cancer Mortality Among Nonsmokers and a Note on Passive Smoking," *Journal of the National Cancer Institute* 66 (1981). *Note:* The author, in a letter published in *The Lancet* on March 3, 1984, was able to show that Garfinkel's American Cancer Society study of passive smoking and lung cancer was consistent with Hirayama's Japanese study, if a correction is made to Garfinkel's study to account for passive smoking at work by 38 percent of his control group.
27. E. C. Hammond and I. J. Selikoff, "Passive Smoking and Lung Cancer with Comments on Two New Papers," *Environmental Research* 24 (1981).
28. "On Passive Smoking," *Medical World News*.

29. U.S. Surgeon General, *The Health Consequences of Smoking—Cancer—A Report of the Surgeon General* (Washington, D.C.: U.S. Dept. of Health and Human Services, 1982).
30. C. T. Irwin, Jr., "Even If You Don't Smoke, It's Bad," *Port Maine Press Herald*, August 13, 1982, p. 1.
31. R. Hutchings, "A Review of the Nature and Extent of Cigarette Advertising in the United States," in Proceedings of the American Cancer Society National Conference on Smoking Or Health, New York, November 18–20, 1981.
32. Ibid.
33. Matthew L. Myers, Craig Iscoe, Carol Jennings, William Lenox, Eleanor Minsky, and Andrew Sacks, *Staff Report on the Cigarette Advertising Investigation*, Public Version (Washington, D.C.: Federal Trade Commission, May 1981).
34. *The Tobacco Observer* 7, no. 1 (February 1982). See also *Cigarette Smoke and the Nonsmoker* (Washington, D.C.: The Tobacco Institute, 1983).
35. W. D. Novelli, "Promoting the Rights of Nonsmokers," in Proceedings of the National Conference on Smoking Or Health, American Cancer Society, New York, November 18–20, 1981, p. 263.
36. *The Tobacco Observer* 7, no. 1 (February 1982).
37. J. L. Repace, "The Problem of Passive Smoking," *Bulletin of the New York Academy of Medicine* 57 (1981).
38. Ibid.
39. A. Brody and B. Brody, *The Legal Rights of Nonsmokers* (New York: Avon Books, 1977).
40. As an example of the propaganda, see "Cigarette Smoke and the Nonsmoker," published by The Tobacco Institute, 1875 I St., N.W., Washington, D.C. 20006 (1983).
41. Roper, "A Study of Public Attitudes."
42. M. Swingle, "The Legal Conflict Between Smokers and Non-Smokers: The Majestic Vice Versus the Right to Breathe Clean Air," *Missouri Law Review* 45 (1980).
43. Public Health Service, State Legislation on Smoking and Health, 1979 (Washington, D.C.: U.S. Dept. of Health and Human Services, 1980); Public Health Service, State Legislation on Smoking and Health, 1980 (Washington, D.C.: U.S. Dept. of Health and Human Services, 1981).
44. C. B. Barad, "Smoking on the Job, the Controversy Heats Up," *Occupational Health and Safety* 48 (1979): 21.
45. Ibid.
46. *Federal Employees for Nonsmokers' Rights, et al. vs. United States of America, et al.*, "Petition for a Writ of Certiorari to the United States Court of Appeals for the District of Columbia Circuit," U.S. Supreme Court, October term, 1979.
47. "Judge Rules Smoke Victim Legally Handicapped," *New York Times*, September 2, 1982.
48. Mike Causey, "The Federal Diary: Landmark Decision for the Nonsmokers," *Washington Post*, October 25, 1982, p. B2.
49. *Shimp* vs. *N.J. Bell Telephone*, N.J. Superior Court Docket no. C-2904-75, December 1976.
50. Brody and Brody, *The Legal Rights of Nonsmokers*; M. Lerner, "Lawsuit Over Smoking Makes Tempers Smoulder," *St. Louis Post-Dispatch*, December 28, 1980, p. 1.
51. Lerner, "Lawsuit Over Smoking."
52. Ibid.
53. *Smith* vs. *Western Electric Co.*, Missouri Court of Appeals, Eastern District, Div. Two, Case no. 44286, September 14, 1982.
54. S. Rovner, "Health Talk: Smoked Out," *Washington Post*, April 4, 1980, p. E5.
55. J. Valente, "Nonsmoking Worker Loses on Appeal," *Washington Post*, May 6, 1983, p. C1.
56. U.S. Environmental Protection Agency, "National Emission Standards for Identifying, Assessing, and Regulating Airborne Substances Posing a Risk of Cancer," App. C, Proposed Rules, *Federal Register* 44, no. 197 (1979), p. 58653.
57. U.S. Surgeon General, *The Health Consequences of Smoking*.
58. G. S. Bonham and R. W. Wilson, "Children's Health in Families with Cigarette Smokers," *American Journal of Public Health* 71 (1981): 290.

59. National Interagency Council on Smoking and Health (NICSH) Business Survey (New York: NICSH, 1980).
60. Repace and Lowrey, "Indoor Air Pollution"; Repace and Lowrey, "Tobacco Smoke, Ventilation, and Indoor Air Quality," *ASHRAE TRANSACTIONS* 88, part 1 (1982): 895.
61. U.S. Environmental Protection Agency, "National Emission Standards."
62. Hirayama, "Nonsmoking Wives."
63. Repace and Lowrey, unpublished manuscript.
64. S. Lichtenstein, P. Slovic, B. Fischhoff, M. Layman, and B. Combs, "Judged Frequency of Lethal Events," *Journal of Experimental Psychology* 4 (1978): 55.
65. The table is adapted from B. D. Dinman, "Occupational Health and the Reality of Risk—An Eternal Dilemma of Tragic Choices," *Journal of Occupational Medicine* 22 (1980); and W. D. Rowe, *An Anatomy of Risk* (New York: John Wiley, 1977).
66. Repace, "The Problem of Passive Smoking."
67. NICSH Business Survey.
68. "Ban on Smoking Makes Good Sense," editorial in the *Seattle Post-Intelligencer*, April 2, 1984.

2
Airborne Contaminants in the Workplace

Laura Punnett

Introduction

The knowledge that hazards to human health can result from working conditions is not new. Bernardino Ramazzini published a description of the diseases common to craftworkers of medieval Europe about the year 1700.[1] In eighteenth-century England, Percival Pott observed that chimney sweeps developed skin cancer of the scrotum because their clothing was constantly permeated with coal soot.[2] In the early 1900s, Dr. Alice Hamilton pioneered in the study of the health effects of exposure to lead and other toxic chemicals in American industry.[3]

Since Hamilton's time, scientific research on occupational diseases has progressed sporadically. The development of a body of knowledge on occupational health hazards is not a simple matter of scientific progress but is marked by frequent political controversy. Workers' movements to improve the quality of their working conditions have been stronger at some times than others in confronting industrial interests concerned with maintaining profit levels and control over the organization of the workplace. National labor strength and other factors such as economic prosperity and social conscience create a political context which is reflected certainly in the extent of government regulation and probably also in the amount and type of scientific research undertaken at different periods of time.

In the late 1960s, a combination of forces brought workplace safety and health issues to national attention: an increase in the reported rate of industrial accidents, rising education and expectations of blue-collar workers, greater awareness of environmental quality, and concern over the potential hazards of the thousands of chemicals produced by the new petrochemical industry following World War II. The Farmington, West Virginia, coal mine cave-in in 1968, which killed 78 miners, led to a massive strike over coal mine safety and provided the final impetus for passage of the Coal Mine Safety and Health Act the next year. This was followed in December 1970 by the Occupational Safety and Health Act. The OSH Act was passed to establish federal mechanisms for standard setting, inspection of workplaces, research into occupational hazards, and educational activities. These functions are accomplished through two discrete agencies. The Occupational Safety and Health Administration (OSHA), in the Department of Labor, was provided with regulatory and enforcement powers. The National Institute for Occupational Safety and Health (NIOSH), under the Department of Health and Human Services, was charged with research and professional education and training of industrial hygienists, epidemiologists, occupational nurses, and physicians.

Since 1971, when the OSH Act took effect, the law and the agencies which it established have provided the context in which almost all occupational health and safety activity takes place in the United States. Other regulatory bodies which also have some jurisdiction in this arena include the Mine Safety and Health Administration, the Federal Railroad Administration, the Nuclear Regulatory Commission, and the National Labor Relations Board. On the state level there are departments of labor and public health, some with their own OSHA plans, and workers' compensation boards.

This chapter will discuss the types of airborne hazards frequently encountered in American workplaces; the difficulties in estimating the impact of these hazards on the health of workers; the methods that are employed or available to reduce the hazards and their adverse effects; and the types of required and recommended standards that exist to guide the reduction of occupational health hazards in the air.

Types of Contaminants

Chemical pollutants in workplace air are generated through a wide variety of industrial and nonindustrial processes and applications and may be present as gases, vapors, mists, fumes, or dusts.[4] Although the presence of contaminants in the workplace environment may appear to be self-evident to workers and employers, in fact this is not always the case. Visible dust or fumes do indicate, at minimum, "nuisance" particulate contamination and may cause general discomfort and irritation if not toxic effects. However, the absence of visible particles does not ensure the absence of toxic substances in the air. Particles small enough to be inhaled (on the order of five microns in diameter) are far too small to be seen by the naked eye. Toxic levels of gases or vapors may not be detected if they have no odor, if they have a high olfactory threshold (they have harmful effects even at concentrations too low to smell), or if employees have developed a tolerance to the odor after chronic exposure. Gases such as carbon monoxide have no warning properties (odor, color, irritative effect) at all.

Often the only way to identify a hazardous substance and its concentration in the air is to perform ambient monitoring for the chemical in question. But air monitoring may not be fully accurate owing to imprecision in the instruments available and variability in emission levels according to production schedules, local weather patterns, and air currents. Monitoring of concentrations in the general work area is less informative than using personal samplers which measure levels in the breathing zone of individual workers.

Biological monitoring is another method that is sometimes relied on to detect the presence of hazardous agents in the workplace. This involves examination or laboratory testing of potentially exposed workers for evidence of accumulation of the substance in body tissues or of early changes in physiological indices. It is controversial whether these early changes should be interpreted as benign indicators of exposure alone or whether, even if potentially reversible, they actually denote the beginning of a disease process. A more important problem is that biological monitoring evades the public health imperative to prevent, as much as possible, all potentially toxic exposures. Rather than use human beings as monitors after the fact, to determine the dose to sensitive body tissues, environmental monitoring should be employed to assess the adequacy of industrial hygiene controls on workplace emissions.

Control Measures

There are several basic approaches to reducing worker exposure to hazardous substances in the air. "Engineering controls" reduce or eliminate the substance at its source. These include enclosure and local exhaust ventilation (laboratory hoods are a good example) and changes in the work process that reduce the amount of contaminant generated or released into the workroom air. Another method is to substitute a less toxic substance, such as the use of silica-free sand in foundries. From the public health point of view, these types of control measures are preferred for their emphasis on primary prevention of emissions into the environment.

Engineering controls are also the most effective type of measure, but only when properly implemented. Overhead or wall fans which ventilate the entire work room don't prevent workers from being exposed to toxic fumes or dusts; they only dilute the concentrations and often emit them into the general environment where community members are exposed. A local ventilation system needs to be designed for the particular contaminant and the workplace in which it will be installed. System maintenance must be regularly scheduled along with air sampling to monitor the system's effectiveness. Clogged ducts or exhaust fans not operating at the proper speed may make the system ineffective despite the illusion of total emission control.

Similarly, substitution of one chemical for another works only if the new agent is indeed less toxic. Since the discovery that benzene causes leukemia, toluene has been substituted in many laboratories and industrial processes. However, commercial toluene is often contaminated with benzene and purchasers do not always require that high standards of chemical purity be met.

Another approach to reducing exposure calls for personal protective equipment to shield workers from airborne chemicals. The kinds of respirators used range from paper dust masks for the nose and mouth to full-face air-supplied hoods. Their effectiveness depends on many factors, including degree of fit to the individual wearer, adequacy of the filter or absorption cartridge for the nature and concentration of the contaminant, and quality of the maintenance program, especially the frequency with which the filter is changed. Respirators can be particularly problematic for any worker whose face does not conform to the model used in designing the facepiece, such as men with beards or women with smaller faces than the "average" man.

Other measures to reduce occupational exposures are characterized as "administrative controls." Some of these, such as education and training of workers, are important adjuncts to any engineering control program. In contrast, however, the rotation of workers in and out of hazardous work areas results in exposing an even larger number of individuals to a toxic substance and therefore is usually a poor substitute for reducing exposure to less hazardous levels.

Health Hazards

Most airborne contaminants enter the body through the respiratory tract. Their health effects may be local, causing damage to the first site of contact, or systemic, if the substance moves to other areas of the body. Thus these agents can cause diseases of virtually every system of the body. The tissues ultimately affected by a particular agent depend on its physical form and size, its route of entry into the body, its distribution and storage patterns, and the sensitivity of different organs to its chemical effects. Health effects that occur within minutes to days after exposure are termed "acute"; "chronic" effects occur months or years later.

Local Effects

Some gases and vapors (sulfur dioxide, chlorine ammonia) are immediately irritating or corrosive to the nose, mouth, and other mucous membranes of the upper respiratory tract, causing coughing, sneezing, and tearing of the eyes. Continual exposure with inflammatory changes may also lead to chronic bronchitis or emphysema. Other gases (ozone, nitrogen oxides) penetrate more deeply into the bronchi and lungs before they begin to cause tissue damage. The effects of these may be fatal if large enough concentrations are inhaled. Inert gases (carbon dioxide, methane) are simple asphyxiants: they displace air in the environment, for example in a closed room, and deprive the lungs of oxygen.

Dusts from minerals such as silica, asbestos, and coal accumulate in the lungs and produce the group of chronic diseases known as "pneumoconioses," meaning the replacement of functional lung with scar tissue. Eventual permanent changes in lung structure and loss of pulmonary function may lead to severe disability and even death, usually from heart failure. Such conditions have historically been extremely prevalent in industries where heavy exposure to these dusts was taken for granted, and even today with knowledge about both their causes and means of preventing them they are still far too common.

Biological dusts are produced from animal and vegetable sources, such as hard woods, animal skin and hair, proteolytic enzymes in detergents, grain in storage silos, and fungal spores from moldy hay or cheese. These dusts, like some synthetic chemicals, can cause either of two diseases. Occupational asthma involves wheezing and shortness of breath continuing for as long as four to six hours after leaving work; it may progress to severe, chronic disease. Allergic alveolitis (inflammation of the small airways) causes fever and flu-type symptoms; long-term exposure may cause lung scarring similar to that of the pneumoconioses. A related disease, byssinosis, is caused by exposure to dust from cotton and similar textile fibers. It appears to be another type of asthma which also gradually progresses to chronic impairment of lung function and may be fatal. (The existence of the chronic form of byssinosis continues to be disputed by some scientists and spokespersons for the cotton textile industry.[5])

Numerous occupational carcinogens affect the respiratory system. Asbestos is a well-documented carcinogen and, due to its ubiquity in the environment, is probably the single largest cause of occupational lung tumors. In addition, the needlelike shape of the asbestos fibers permits them to penetrate the lining of the chest and abdominal cavity to cause mesothelioma, another type of cancer. The emissions from coking ovens in steel mills are mixtures of particulates, gases, and vapors, including a family of highly carcinogenic hydrocarbons. Long-term workers on the topside of these ovens, who have been predominantly black men, had a rate of lung cancer ten times greater than that of the general population before engineering controls were instituted.[6] Naturally occurring radioactive gases, which diffuse from brick, granite, and rock into the surrounding air and water, become concentrated in the air in uranium and other mine shafts. There is evidence that uranium miners (Native American and white) in the southwestern United States suffer elevated rates of lung cancer not attributable to other causes.[7]

Airborne particles may also be swallowed, either because they are too large to be respirable and are cleared from the nasal passages to the throat, or as a result of eating, drinking, or smoking in contaminated workplace air. Ingestion of asbestos fibers in this way is believed to be the cause of some stomach and intestinal cancers among asbestos workers.

Systemic Effects

Some substances are absorbed from the lungs or through the skin into the bloodstream and transported to other organs of the body. They may then cause damage to various sensitive tissues (target organs) as well as being stored in other tissues, metabolized into more or less toxic forms, or excreted. Any organ system may be damaged by airborne environmental toxins, although certain sites in the body tend to be more commonly affected because of common routes of distribution and excretion.

Lead is an example of an airborne contaminant which affects organs throughout the body. Inhaled in the form of fume or dust particles, it is stored in the bones but is toxic to the brain and nervous system, the blood-forming tissues, the digestive tract, and even to the kidneys as they excrete it. Lead is also a reproductive hazard, in that it may cause impotence and sperm abnormalities in men and is toxic to the developing embryo or fetus.

The liver is the organ which metabolizes and detoxifies many substances, and so it is vulnerable to their harmful effects. Examples include fat-soluble chemicals such as PCBs (polychlorinated biphenyls, which are also a major environmental contaminant in some parts of the United States), many pesticides, organic solvents such as carbon tetrachloride, and other aromatic or chlorinated hydrocarbons. (Organic solvents are used in laboratories, paints, degreasing and cleaning operations, and other manufacturing processes.) Arsenic, phosphorus, and copper salts are toxic to the cells of the liver; dusts and mists emitted in the manufacture of oral contraceptives and other pharmaceutical drugs affect liver functioning as well as causing direct cellular damage. A number of chemicals to which workers receive airborne exposures are also liver carcinogens. Vinyl chloride causes a rare form of cancer known as angiosarcoma of the liver. Exposure to some pesticides, including dioxin, and to solvents such as trichloroethylene and carbon tetrachloride is suspected of increasing workers' risk of liver cancer.

The kidney and bladder, because of their role in detoxification and excretion, also concentrate toxic substances and suffer chemical injury as a result. Many metals, such as cadmium, mercury, and lead, cause kidney damage. Arsine gas (produced in the chemical, smelting, and metal refining industries) is extremely toxic to the kidney, fatally so in high concentrations. Aromatic amines are well-known carcinogens; high rates of bladder cancer were documented in the 1950s in dye workers exposed to benzidine and napthylamine, although such exposures have presumably been rare since those findings were widely disseminated.[8]

Certain chemicals have effects on the hematopoietic (blood-forming) system or on already-formed blood cells. Lead disrupts the synthesis of hemoglobin, which is necessary for the delivery of oxygen to body tissues. Carbon monoxide interferes with hemoglobin transport of oxygen, causing asphyxia. Arsine causes widespread rupture of red blood cells (hemolysis), as do lead and TNT, although their effect is less potent. Benzene interferes with the production of white blood cells (necessary for immune response) in the bone marrow, causing aplastic anemia and eventually leukemia. Ethylene oxide, a sterilant gas used in hospitals, clinical laboratories, and dental clinics, has also been linked with leukemia.

Other chemicals affect the heart and circulatory system directly. Chloroform vapors and fluorocarbon gases produce cardiac arrhythmia and may cause heart attacks. Chronic exposure to carbon disulfide (a gas emitted in viscose rayon manufacture and cold curing of rubber) leads to hardening of the arteries and coronary heart disease. Kidney damage due to metal dusts such as lead and cadmium often leads to secondary hypertension and heart disease.

Reproductive hazards include any substance which interferes with the normal function of the male or female reproductive system or is toxic to the developing embryo or fetus. A number of different chemicals are damaging to the endocrine glands involved in the reproductive system, with effects such as menstrual disorders in women and impotency in men. For example, pesticides such as Kepone® and DBCP are toxic to the testicles, suppressing sperm production and leading to male infertility. Lead, cadmium, and organic tin compounds have been shown to have similar effects in animals. Pharmaceuticals such as steroids, cytotoxic drugs, and other endocrinologically active agents are toxic to the adrenal glands.

Other chemicals, labeled teratogens, interfere with the normal growth and development of the embryo and fetus. Lead and mercury are examples of occupational pollutants which can be absorbed into the maternal bloodstream and transported across the placenta to the fetus, where they may induce central nervous system damage and congenital neuropsychological disease respectively. The prostate gland has the unique ability to concentrate and excrete heavy metals, such as zinc and cadmium, into the seminal fluid; this could also pose a source of toxic exposure to the developing embryo or fetus. Anesthetic gases cause miscarriages and birth defects in offspring of women exposed occupationally (workers in surgical and dental operating rooms) as well as female partners of exposed men.

Children of exposed workers are sometimes at risk for toxic effects after birth. Fat-soluble compounds will accumulate in breast milk and are passed on to a nursing infant. There is conflicting evidence of elevated rates of cancer in children of men occupationally exposed to hydrocarbons (solvents in paints and cleaners; oils, gasoline, and other petroleum fuels; etc.).[9] Such an effect could be an example of mutations in the sperm of exposed male workers, causing genetic changes that can be transmitted from one generation to the next.

The nervous system, both central and peripheral, is affected by many common workplace exposures. Peripheral neuropathy involves loss of motor and sensory function (weakness, tingling, or numbness) in the limbs and may be irreversible if there is long-term exposure and progression of the syndrome. It results from exposure to lead and to vapors of the solvents n-hexane and methyl butyl ketone (even as a small proportion of solvent mixtures). Certain pesticides interfere with the autonomic nervous system to cause headache, contraction of the pupils, sweating, muscle twitching, gastrointestinal pain, and vomiting.

Central nervous system disorders may range from behavioral changes, including mild fatigue, mood changes, and short-term memory loss, to toxic psychoses. These effects can be induced by many different chemicals, even at low doses, such as lead, mercury, arsenic, carbon monoxide, carbon disulfide, organophosphate pesticides, and a variety of solvents. Overt toxic psychosis is rare at the lower exposure concentrations seen now in industrialized countries, but is known to be caused by arsenic, lead, manganese, mercury, and carbon disulfide. When solvents and anesthetic gases act on the brain and central nervous system, they also impair reflexes, mental alertness, and judgment. Such impairment may lead to accidents and cause injury to the affected worker as well as other people in the workplace.

Brain cancer, although uncommon in the general population, is associated with occupational exposure to vinyl chloride. It has also been found at rates higher than expected among workers in the petrochemical and rubber industries, although the precise cause is still unknown.

Estimates of Occupational Exposure and Disease Rates

For a number of chemicals known to be toxic or carcinogenic, estimates have been made of the prevalance of exposure by industry and occupation. Table 2.1 lists 30 hazardous substances found in American workplaces with estimates of the number of workers exposed from data compiled and published by NIOSH in the 1970s. The figures presented are a summation across a wide variety of work settings. For example, carbon monoxide is a common by-product of many different industrial operations; asbestos and benzene exposures occur during manufacture of the primary product and also as a result of many different uses of the product under a broad range of conditions. These estimates necessarily suffer from inaccuracy both because of changes in industrial processes and formulations and because of fluctuations in the economy and employment levels. In addition, it is not always clear whether "exposure" includes occasional work with a substance regardless of whether exposure actually occurred, or whether it is restricted only to known regular or frequent work exposures.

It should also be kept in mind that these estimates do not include indirect exposures, such as those to family members when an employee wears or carries contaminated work clothing to the home. This is a common source of exposure to asbestos fibers, with their associated cancer hazard, and to lead dust, which poses the risk of irreversible neuropsychological damage to the children of lead-exposed workers. Farm-worker families, who often live immediately next to cultivated fields, are exposed to pesticides not only while at work but also when chemical solutions are sprayed on the fields and drift with air currents into their living areas.

The surveys that NIOSH has undertaken have, for the most part, covered only industrial hazards and those known for many years to be toxic. Exposures which occur in offices (vapors from duplicating solvents, correction fluids, and cleaning solutions), laboratories and hospitals (anesthetic gases, sterilizing gases, and pharmaceutical drugs), and other nonindustrial workplaces have been overlooked. One consequence is that the magnitude of the health hazards in many traditionally female occupations is even more poorly estimated than in the manufacturing sector.

Although there are problems inherent in determining the number of workers exposed to different substances, it is even more difficult to assess the magnitude of disease caused by these exposures. There exists no national registry or reporting mechanism for collecting data on new cases of occupational disease. One direct way of estimating these numbers would be to multiply exposure frequencies by the rates of disease in the exposed populations. However, there are few substances for which we have pre-

Table 2.1 Estimates of Numbers of People Occupationally Exposed to Selected Chemicals and Other Airborne Substances

Substance	Number Exposed (in Thousands)
Aromatic hydrocarbons	3897
Arsenic	1500[a]
Asbestos	200[a]
	1565
Benzene	1919
Cadmium	100[a]
Carbon dioxide	2963
Carbon disulfide	1088
Carbon monoxide	2000[a]
Carbon tetrachloride	160[a]
	2007
Chloroform	40[b]
	80[b]
Chromium (hexavalent)	75[a]
Cotton dust	230[a]
	800[b]
	525
Cutting oils	1229
Ethylene dibromide (EDB)	9[b]
Fiber glass	200[a]
Formaldehyde	1755
Grain dust	500
Hexane	2053
Isocyanates, toluene (TDI)	40[b]
Lead (inorganic)	83[a]
	1418
Mercury	150[a]
Nitric oxide	350[a]
	2016
Organophosphate pesticides	845
Polychlorinated biphenyls (PCB's)	12[b]
Silica, crystalline	1200[a]
	3213
Sulfur dioxide	500[a]

(Continued)

Table 2.1 (Continued)

Substance	Number Exposed (in Thousands)
Toluene	100[a]
	4811
Trichloroethylene (TCE)	200[a]
	3652
Vinyl chloride	2267
Wood dusts	1646

a. The President's Report on Occupational Safety and Health (Washington, DC: GPO, 1973).
b. NIOSH Alerts, Bulletins, Letters, or Criteria Documents, as cited in Barth, note 10.
Sources: National Occupational Hazard Survey, vol. III, Survey Analysis and Supplemental Tables (DHEW (NIOSH) Pub. #78-114, 1977), unless otherwise noted.

cise estimates of the rates of disease they cause. Even those disease rates that have been estimated cannot be indiscriminately applied from one exposed population to another. This is because exposure is defined differently in different studies, as mentioned above, and also because actual exposure concentrations may vary greatly among plants due to their age and physical condition, the type of process involved, whether emission controls are in place, other substances present in the air simultaneously, and so on.

Another method is to apportion, somewhat arbitrarily, the percentages of various diseases which are due respectively to diet, cigarette smoking, environmental pollution, occupational hazards, etc. These percentages are then applied to the number of deaths in the general population annually from each disease. By this means, several different authors have independently arrived at figures in the range of 500,000 new cases of occupational disease and 100,000 to 200,000 occupational disease fatalities each year in the United States.[10] There is no way to objectively assess the validity of these estimates nor to know what proportion of these should be ascribed to airborne hazards, although certainly a fairly large proportion of occupational health hazards are in respirable form.

A third approach has been to rely on statistics from employers' reports to OSHA, workers' compensation systems, or other state-administered reporting mechanisms for occupational disease. The use of data from each of these sources is problematic and often leads to underestimation of true incidence and mortality rates. The sources of bias lie in reasons related to the definition of occupational disease and the actual mechanism by which an individual's disease might (or might not) be diagnosed or compensated as work-related. A few of the major issues are discussed below.

Relatively few diseases are truly occupation-specific, occurring only in workers exposed to a particular toxic substance. The vast majority of occupational diseases also occur as a result of nonoccupational causes or may be caused or aggravated by interaction between workplace contaminants and nonoccupational factors, such as cigarette smoking or environmental air pollution. Metal-fume fever, which affects welders and smelter workers exposed to zinc, copper, or magnesium fumes, produces fatigue, muscle ache, fever, and general malaise quite similar to the flu. Chronic industrial bronchitis is found among miners, foundry and smelter workers, and rubber and fertilizer manufacturers. It is also virtually indistinguishable from the bronchitis affecting long-term smokers. Cirrhosis of the liver caused by industrial chemicals follows precisely the model of cirrhosis caused by excessive alcohol consumption. Stomach or lung cancer in an asbestos worker cannot be readily distinguished from malignancy in an unexposed individual. Therefore, in order for most disease to be recognized and reported as work related, there must be evidence that exposure has been sustained *and* knowledge that the exposure is in fact toxic.

The right of workers to obtain information about the substances they work with is crucial both to prevention and to compensation. Many people still have little or no information about whether they are exposed to any hazards in the course of their employment. The labor-management dynamics in most workplaces coupled with job insecurity generally function to discourage workers from obtaining information on their past or present exposures which might be relevant to present or future illness.

Legal guarantees of the right to the names of substances and their health effects have been slow and inadequate to date. The labeling standard proposed by OSHA in 1980 was subsequently withdrawn by the Reagan administration. A new hazard communication standard was proposed in 1983, to take full effect by 1986. This standard is significantly weaker in several respects; it covers only employees in the chemical manufacturing sector and has no requirements for labeling of pipes or transfer containers used in the production line (the only ones that many exposed workers would see).[11]

Some states, counties, and municipalities have passed "Right-to-Know" statutes as a result of lobbying efforts by labor, community, and environmental groups; these regulations tend to be more stringent than the proposed federal standard. For example, the Massachusetts law (1983) states that the Department of Public Health will determine which chemicals are covered and the information the labels must provide. Under the OSHA standard, chemical manufacturers would be able to make that determination. Nevertheless, OSHA intends that the hazard communication standard will preempt state legislation. This move will probably be challenged in the courts, since the OSH Act does permit the states to set more stringent standards than the federal government if they choose.

Even when information is available to workers about the substances they have worked with, there must be an involved party sufficiently knowledgeable to link the past exposures with current disease. One might expect that this person would be either a personal health-care provider or a company-employed physician or nurse. Unfortunately, many providers have received little or no training in occupational medicine. They may fail to inquire about chemical exposures at work, or they may not know the importance of such information if it is offered.

The relation between occupational exposure and illness might also be obscured for two other reasons: long latency period (the delay between first exposure and manifestation of health effects), and multiple or interacting causes of disease. Many chronic occupational and other environmental health hazards have latency periods as long as 10 to 40 years before they produce clinical disease. A worker might hold other jobs in the interim between exposures and disease; or, in the same employment, processes and materials may change so that recent exposures are not informative about past hazards. Thus the relevant exposures may never be inquired after by the medical provider or offered by the person seeking treatment. Furthermore, if a disease is not manifested until after retirement, it may never become known to the employer and will almost certainly not be reported on the OSHA log of recordable injuries and illnesses, which covers only current employees.

The problems inherent in deciding whether an illness is occupational in origin are greatly compounded when a disease cannot be definitively attributed to a single cause. If the exact causes and mechanisms of the disease are not known, as with many psychiatric disorders, for example, it is virtually impossible to assess the contribution of the workplace. Or a disease may be known to have multiple risk factors, as most chronic illnesses such as cancer and heart and lung diseases do, in which case the decision of how much to attribute to occupational factors is rarely easy and often somewhat arbitrary. The scientific data on this question tend to be inconclusive and capable of being interpreted in various ways to justify more than one point of view. For example, some authors ascribe most lung cancer incidence to cigarette smoking, despite evidence of interaction between tobacco and various occupational hazards.[12]

An additional set of problems with data from workers' compensation systems arises from the fact that these systems were set up at the beginning of this century to deal primarily with injuries rather than illnesses, and they retain severe inadequacies in their compensation of occupational disease. Since occupational disease is compensated according to the extent of disability, impairment which does not disable a person relative to his or her current job is not compensated as a rule.

Furthermore, company medical providers may see it as part of their job to keep compensation costs down by identifying nonoccupational causes of the illness. Since companies and their insurance carriers have an incentive to contest a claim proportional to its amount, diseases which cause permanent and total disability are most likely to be contested. This leads to their vast under-representation in compensation data relative to non-disabling conditions.[13] And since each case must be judged individually, without reference to epidemiologic evidence or previous settlements of similar cases, the exposure-disease link must be reestablished each time. This places a difficult burden of proof on the individual, even when it is clear epidemiologically that workers with that exposure suffer a rate of that illness which is much higher than in the general population. Barth has shown that, even in states with relatively progressive compensation systems, as few as one percent or less of occupational disease fatalities per year are compensated.[14]

Death certificates are another source of data for epidemiologic research on causes of mortality. These are filled out by a variety of personnel (physicians, coroners) without a standardized procedure to follow. As a result, they often note only the most recent or usual occupation for a person actively employed at time of death. Often no job information is recorded for retired workers, and women may be coded as "housewives" even if they also worked outside the home.

Legal and Contractual Remedies

The concentrations of airborne contaminants in American workplaces have for the most part decreased greatly since the earlier part of this century. This improvement has been in response to pressure in different periods of time from the public health sector, labor, government, and public opinion; avoidance of legal liability for effects on human health may also be a frequent motive. Most substances have been eliminated or controlled voluntarily, in accordance with recommendations from the scientific community, or as an indirect result of changes in manufacturing processes. A few substances have seen a more rapid cleanup since the promulgation of specific standards by OSHA, especially in larger industrial workplaces, where enforcement tends to be more consistent.

Other hazardous substances have been newly introduced into the workplace, however, or have become more ubiquitous with increasing production of petrochemicals and their derivatives, such as pharmaceuticals. Byssinosis has actually become more prevalent as a result of technological changes such as mechanization of cotton picking and the development of high-speed carding machines, both of which generate more cotton dust and debris in the mills.[15]

There are several sources of recommendations and standards regarding levels of workplace emissions. The American Conference of Governmental Industrial Hygienists (ACGIH), with representatives of private industry, government, and academia, publishes voluntary guidelines known as "threshold limit values." This list covers over 650 different chemicals and physical agents which are found as airborne contaminants; the levels are average concentrations which are intended to be safe for continuous exposure over a 40-hour work week.[16]

Threshold limit values are set by consensus, based on data ranging from animal toxicology to case reports of acute industrial exposures. With the exception of about 50 recognized or suspected carcinogens, many of the levels are set to prevent acute health effects rather than chronic. But as the levels of many toxic substances have decreased over time, acutely fatal exposures have become less common and there is a greater need for accurate information on chronic effects of exposures to lower concentrations. While it is true that these data are frequently incomplete or inconclusive, it is unnecessary to require rigorous scientific proof before preventive action is taken. "An appropriate public health orientation dictates . . . serious efforts aimed at preventing and controlling occupational disease, even in the absence of definitive scientific knowledge" on such matters as causal mechanisms or dose-response curves.[17]

NIOSH is responsible for the continuing evaluation of existing standards and recommended exposure levels. The institute collects and publishes data on current workplace exposures, both short and long term, and conducts its own studies on plants and industries where it has reason to believe that a health hazard may exist. NIOSH then recommends to OSHA that new standards be set or existing standards be modified to reflect new findings.

OSHA is the federal agency with legal authority to enforce compliance with legally mandated exposure levels in every nongovernmental workplace with ten or more employees. Since its establishment, the agency has promulgated a number of standards which have the force of law. Many of these regulate specific chemicals or mixtures, such as asbestos, vinyl chloride, arsenic, lead, coke oven emissions, and cotton dust. Each standard defines not only permissible exposure levels but other, equally important procedures such as environmental monitoring, the use of respirators and other personal protection, labeling of containers, and medical surveillance for occupational disease in exposed workers.

The original intent of the OSH Act was that these standards would constitute the primary means of regulating the vast majority of hazardous substances found in American workplaces. However, the standard-setting process has proved to be quite adversarial in nature and extraordinarily time consuming as a result of both administrative procedures and challenges in the courts. As a result, fewer than ten health standards on chemical exposures have been successfully written into law, in contrast to the tens of thousands of toxic chemicals used in both industrial and nonindustrial workplaces.[18]

The protection afforded by those health standards which are in existence is also limited. OSHA standards are arrived at through a process of political compromise between the mandate to protect workers' health to the greatest extent possible and technological and economic feasibility for the industries involved. The act mandates that permissible exposure levels for a toxic substance shall be established which ensure "that no employee will suffer material impairment of health or functional capacity" as a result of exposure even for his or her entire working life. To set such a level requires certain knowledge about questions such as whether a safe threshold exposure exists or how to extrapolate dose-response curves from higher to lower doses. Yet these are usually areas of great scientific uncertainty even when a hazard is clearly serious enough to warrant prompt action to minimize further exposures.

The reality has been that in the process of setting each standard, industry's statements about what control measures are technically possible and the projected costs of those controls have been weighed against the projected health benefits which would accrue to workers by reducing exposures. In its 1981 decision upholding the cotton dust standard as promulgated by OSHA, the U.S. Supreme Court held that OSHA was not required to and in fact should not engage in formal cost-benefit analysis in setting health standards. However, risk assessment is used in standard setting and has been required of the agency since the Supreme Court's rejection of the benzene standard in 1980.[19]

In public hearings for proposed health standards, individual workers and their organizations do give testimony on the impact of occupational hazards on their health and quality of life. However, the workers themselves, as the people most directly affected by OSHA standards, do not participate in the final decision on what level of risk is "acceptable." The weighing of costs to industry against benefits to employees ultimately takes place informally within OSHA, although a standard once promulgated can be, and usually is, appealed in the courts, sometimes by both industry and labor separately.

Unlike the ACGIH threshold limit values, at least some of which are reviewed for updating every year, OSHA standards can easily become outdated as new scientific evidence becomes available. For example, the asbestos standard currently in effect establishes a permissible exposure level intended to protect against chronic lung disease (asbestosis), but it does not protect against cancer nor does it address the question of whether a safe threshold level exists for any carcinogen. Theoretically OSHA standards may be updated, but the entire administrative process of notice, hearing, promulgation, and usually appeal must be repeated as if it were a new standard. Given the number of chemicals still to be regulated and other demands on its time and resources, the agency has not yet issued a second, revised standard for any substance.[20]

Toxic chemicals are often studied in animals or in human populations in isolation, in order to determine their effects independent of any other factors. But workers are often exposed to mixtures in the air rather than to single substances, and it may be difficult to determine which of these is directly responsible for any adverse effects. In fact, frequently these agents do not act in isolation but rather interact synergistically in more or less complex ways. For example, when particulate matter such as carbon dust is present in the air simultaneously with sulfur dioxide gas, the gas adsorbs onto the particles and is carried farther down into the respiratory tract than it would penetrate by itself. This leads to chemical irritation deeper in the lung, with potentially more serious effects. Should sulfur dioxide and respirable dusts each be regulated to lower levels when they are both in the air than when only one is being emitted? Sulfur dioxide is also suspected of being a cocarcinogen, that is, of promoting the effects of carcinogens such as arsenic. Should sulfur dioxide be reduced to the lowest possible level, to prevent carcinogenic effects, or should only arsenic and other "initiators" be regulated as carcinogens? To date OSHA has not grappled with these controversies in its regulatory process; each substance has been treated independently, as if with an implicit assumption that the effects of simultaneous exposures need not be taken into account.

OSHA has alternatives other than setting permissible exposure levels for individual chemicals or mixtures. The coke oven standard sets a maximum exposure concentration for one fraction of the emissions mixture. In addition, specific engineering controls and work practices are required which had been shown previously to reduce emissions from the ovens. Standards could be promulgated without setting levels for any substances, but instead with specifications of control measures which in OSHA's opinion would be most effective in reducing exposure levels.

The other mechanism under the act for enforcing improvements in workplace conditions is the "general duty clause," which specifies that employers must provide working conditions "free from recognized hazards that . . . cause death or serious physical harm." The clause was intended to be used as a last resort, in the absence of a standard, rather than as a substitute for promulgating new regulations.

Neither the act nor its legislative history defined the terms "recognized" or "serious physical harm," and their interpretation has evolved through numerous court challenges. For example, a substance that is shown conclusively by epidemiologic or toxicologic research to be toxic could be covered by the general duty clause. Furthermore, if one manufacturer has eliminated a certain hazard from the workplace, other manufacturers and even other industries could be held to the same standard. The general duty clause was even used, unsuccessfully, to cite a chemical company for an employment policy requiring women of reproductive age to be sterilized if they wished to work in lead-exposed areas.

Employers have protested that citations issued under this authority are arbitrary and lacking in due process, since it is impossible to predict what OSHA would consider a violation and how to comply in advance of a citation. However, with so few specific standards in existence, OSHA in the past has felt that the general duty clause was a necessary and justifiable vehicle, noting that it has been invoked only for hazards about which a scientific or policy consensus existed. In the present antiregulatory climate (1984), OSHA is more likely to use its discretion *not* to find violations in the absence of specific standards.

Another source of standards is the individual states, which are authorized to set permissible exposure levels lower than those required by OSHA for workplaces within that state. Although this authority is rarely, if ever, utilized, some states such as Massachusetts and Pennsylvania have published their own recommendations for a large number of toxic chemicals.

Employees have certain other resources when faced with airborne health hazards in the workplace. Organized workers can bargain for specified work practices, engineering controls, and maximum exposure concentrations in their union contracts, thus obtaining controls over hazards not covered by law or more stringent measures than required by OSHA for regulated chemicals. The basis for such contract clauses might be in ACGIH, NIOSH, or independent scientific recommendations. Both organized and unorganized workers have used strikes and other job actions to pressure employers into improving health conditions. The success of this avenue obviously depends on the immediacy of the hazard, the strength of the labor union, the extent of economic and job insecurity, and the depth of employer concern for health and safety in the workplace.

Both the OSHA Act and the National Labor Relations Act have defined certain situations in which workers as individuals or as a group may refuse to perform a job which they feel puts them in "imminent danger."[21] The right to refuse hazardous work under either act is sharply limited to particular conditions under which that refusal might take place. Many observers feel that the laws are therefore of little benefit to most employees, who are not familiar in advance with the details of the procedures they must follow when confronted with a hazardous task. Even in those situations where a worker's right to refuse has ultimately been upheld in a grievance hearing or in the courts, only imminent safety hazards (danger of immediate injury or death) have been covered. The right to refuse exposure to health hazards, even to a well-documented carcinogen such as asbestos, has not been successfully tested, so it would be misleading to assume that legal protection has been extended in fact to the threat of chronic disease.

It is even more problematic what recourse is available to female employees who face discriminatory employment practices as a result of reproductive hazards present in the workplace. Employment policies in increasing numbers of companies prohibit the hiring of women of childbearing capacity in work areas where teratogens are present. The rationale is to prevent harm to the fetus that would occur if a worker became pregnant while exposed. However, workplaces that traditionally rely heavily on female labor, such as hospitals or electronics plants, have not instituted exclusionary policies even though there may be extensive exposure to reproductive hazards (anesthetic gases, benzene, ethylene oxide, etc.). Nor have men been excluded from employment which exposes them to mutagenic or teratogenic agents, such as lead. If protection of human health were the genuine motivation for these policies, one would expect to see a high priority on the reduction of airborne concentrations to levels safe for all workers. Instead, the suspicion is raised that industry's real concern is to prevent future lawsuits by offspring with congenital defects. One consequence of exclusionary policies is that women are denied jobs that represent opportunities for better wages and benefits than are available in traditionally female occupations. The legal rights of women denied employment under these policies are uncertain. The OSH Act does not cover them explicitly; the Equal Employment Opportunity Commission is probably the best recourse, while union contracts and grievance procedures offer another avenue for fighting these cases.

Notes

This chapter has benefited from discussions with Mary White, Charles Levenstein, and the Women's Committee of the Massachusetts Coalition for Occupational Safety and Health.

1. Bernardo Ramazzini, *The Diseases of Workers (De Morbis Artificum)* (New York: Hefner Publishing Company, 1964).

2. Joseph K. Wagoner, "Occupational Carcinogenesis: The Two Hundred Years Since Percival Pott," *Annals of the New York Academy of Science* 271 (1976): 1-4.

3. Alice Hamilton, *Industrial Poisons in the United States* (New York: Macmillan Publishing Co., 1925).

4. A thorough catalog of these hazards is beyond the scope of this chapter. For more detailed information see the list of references.

5. For a summary of the evidence and arguments see D. H. Wegman, C. Levenstein, and I. A. Greaves, "Byssinosis: A Role for Public Health in the Face of Scientific Uncertainty," *American Journal of Public Health* 73, no. 2 (1983): 188-92.

6. J. William Lloyd, "Long-Term Mortality Study of Steelworkers, V., Respiratory Cancer in Coke Plant Workers," *Journal of Occupational Medicine* 13, no. 2 (1971): 53-68.

7. V. E. Archer, J. K. Wagoner, and F. E. Lundin, "Lung Cancer Among Uranium Miners in the United States," *Health Physics* 25 (1973): 351-71.

8. R. A. M. Case, M. E. Hosker, D. B. McDonald, and J. T. Pearson, "Tumours of the Urinary Bladder in Workmen Engaged in the Manufacture and Use of Certain Dyestuff Intermediates in the British Chemical Industry, Part I. The Role of Aniline, Benzidine, Alpha-Naphthylamine, and Beta-Naphthylamine," *British Journal of Industrial Medicine* 11 (1954): 75-104.

9. An association between childhood cancer and fathers employed in occupations exposed to hydrocarbons was reported by Jacqueline Fabia and Truong Dam Thuy, "Occupation of Father at Time of Birth of Children Dying of Malignant Diseases," *British Journal of Preventive and Social Medicine* 28 (1974): 98-100. Evidence refuting this hypothesis is found in M. Zack, S. Cannon, D. Loyd, C. W. Heath, J. M. Falletta, B. Jones, J. Housworth, and S. Crowley, "Cancer in Children of Parents Exposed to Hydrocarbon-Related Industries and Occupations," *American Journal of Epidemiology* 111, no. 3 (1980): 329-36.

10. Peter S. Barth with H. Allan Hunt, *Workers' Compensation and Work-Related Illnesses and Diseases* (Cambridge, Mass.: The MIT Press, 1980).

11. *Federal Register* 48, no. 228 (Nov. 25, 1983): 53340-47.

12. A well-known and highly regarded source which attributes lung cancer almost exclusively to cigarette smoking is Richard Peto, "The Causes of Cancer: Quantitative Estimates of Avoidable Risks of Cancer in America Today," *Journal of the National Cancer Institute* 66 (1981): 1091-219. Evidence of interaction between smoking and other exposures is provided, among others, by Irving J. Selikoff, E. Cuyler Hammond, and Jacob Churg, "Asbestos Exposure, Smoking, and Neoplasia," *Journal of the American Medical Association* 204 (1968): 106-12; and Alice S. Whittemore and Alex McMillan, "Lung Cancer Mortality Among U.S. Uranium Miners: A Reappraisal," *Journal of the National Cancer Institute* 71, no. 3 (1983): 489-99.

13. Barth, *Workers' Compensation*.

14. Peter S. Barth, "The Effort to Rehabilitate Workers' Compensation," *American Journal of Public Health* 66, no. 6 (1976): 553-57.

15. Jacqueline K. Corn, "Byssinosis: An Historical Perspective," *American Journal of Industrial Medicine* 2 (1981): 331-52.

16. American Conference of Governmental Industrial Hygienists, *Threshold Limit Values for Chemical Substances and Physical Agents in the Work Environment with Intended Changes for 1983-84* (Cincinnati: ACGIH, 1983).

17. Wegman, Levenstein, and Greaves, "Byssinosis: A Role for Public Health."

18. Between 10 and 20 more health standards on chemicals have been proposed but never promulgated as final standards, or were promulgated and thrown out by the courts before being enforced.

19. *American Textile Manufacturers Institute, Inc. v. Donovan, Secretary of Labor,* 101 S. Ct. 2478 (1981); *Industrial Union Department v. American Petroleum Institute,* 448 U.S. 607 (1980).

20. As of March 1984, an emergency temporary standard which would have lowered the permissible exposure level for asbestos has been rejected by the U.S. Court of Appeals for the Fifth Circuit, and a new permanent standard is expected to be published shortly.

21. The Supreme Court decision which upheld this right under OSHA was *Whirlpool Corporation v. Marshall, Secretary of Labor,* U.S. Law Week, 1980: 48: 4189-95.

References

Burgess, William A. *Recognition of Health Hazards in Industry: A Review of Materials and Processes.* New York: John Wiley and Sons, 1981.

Hricko, Andrea, with Brunt, Melanie. *Working for Your Life: A Woman's Guide to Job Health Hazards.* Berkeley, Calif.: Labor Occupational Health Program, 1976.

Hunt, Vilma. *Work and the Health of Women.* Boca Raton, Fl.: CRC Press, 1979.

Key, Marcus M.; Henschel, Austin F.; Butler, Jack; Ligo, Robert N.; and Tabershaw, Irving R., eds. *Occupational Diseases: A Guide to Their Recognition.* Washington, DC: DHEW(NIOSH) #77-181, 1978.

Levy, Barry S., and Wegman, David H., eds. *Occupational Health: Recognizing and Preventing Work-Related Disease.* Boston: Little, Brown, 1983.

Schilling, R. S. F., ed. *Occupational Health Practice.* 2nd ed. London: Butterworths & Company, Ltd., 1981.

Stellman, Jeanne M., and Daum, Susan M. *Work Is Dangerous to Your Health.* New York: Random House, Inc., 1973.

3

The Risks from Transported Air Pollutants

Robert M. Friedman

Introduction

Acid deposition, commonly referred to as "acid rain," has been the focus of much debate during the Clean Air Act deliberations of the 97th and 98th Congresses. The controversy over transported air pollutants—of which acid rain is one example—revolves around potential adverse effects on lakes and streams, forests and soils, agricultural crops, building materials, and human health. Concerned about the potential magnitude of these damages, some individuals and groups have advocated that Congress take more stringent measures to control pollutant emissions. Others, pointing to the uncertainties in our knowledge of the causes and consequences of transported air pollutants, are concerned that the costs of such actions may be unwarranted.

Reducing total sulfur dioxide emissions—the major source of acid deposition in the eastern United States—by more than about 25 percent is estimated to cost billions of dollars annually. Efforts to control transported air pollutants could also significantly affect the coal industry. Stringent sulfur dioxide emissions controls could lower the demand for high-sulfur coal, displacing mining jobs and disrupting mining communities.

In addition to concerns about resource damage and costs of control, transported air pollutants may impose significant inequities, as they affect different groups, regions, and even nations from those receiving the benefits of the activities that generate acid rain and ozone. Moreover, groups and regions may bear an unequal share of the costs of transported air pollutant control.

Recognizing that the uncertainty surrounding many aspects of the problem complicates the decision of whether, or when, to control transported air pollutants, the committees responsible for reauthorizing the Clean Air Act asked the Office of Technology Assessment to evaluate the potential implications of such transported air pollutants as acid deposition and ozone. In response to the concerns of Congress, OTA has examined the potential benefits of acting now to abate transported air pollutants versus the potential costs of action that may be premature. This chapter is based upon that report.[1] Unpublished in-house or OTA-sponsored work provides the basis for the estimates presented below; however, the views expressed in this chapter are those of the author and not necessarily those of OTA.[2]

Current understanding of transported pollutants allows a description of potential resource damage—regions of the United States most susceptible to damage—but not a precise quantification. The costs of likely control strategies, and who might bear these costs, can be described, but the effectiveness of these control strategies for avoiding resource damage cannot be calculated with confidence. Given the difficulty of quantifying the relationship between emissions reductions and the prevention of resource damage, near-term policy decisions must be based on the risks of resource damage, the risks of unwarranted control expenditures, and the distribution of these risks among different groups and regions of the country.

Transported Air Pollutants

Air pollutants travel varying distances; in some cases, hundreds or even thousands of kilometers. "Local" transport is pollution originating from a source within a distance of 50 km (about 30 miles); "long-range" transport refers to distances of about 300 km to 2000 km. Measured pollutant levels in any given location are the cumulative result of local, medium, and long-range transport.

Acid deposition and ozone can both be transported long distances. They result from the chemical transformation of three pollutants: sulfur dioxide, nitrogen oxides, and hydrocarbons. "Acid rain" is commonly used to refer to a variety of pollutants deposited in both wet and dry forms. Acid deposition—the term that more accurately represents the scope of the phenomena—results from gaseous sulfur and nitrogen oxides, sulfates and nitrates (transformation products of these gases), and from other chemicals in the atmosphere. Acid deposition is a combination of directly emitted or "primary" pollutants and transformed or "secondary" pollutants.

The secondary pollutants—sulfates, nitrates, and ozone—can also manifest themselves in other ways. Concentrations of airborne sulfate and nitrate particles are correlated with reductions in visibility. Sulfate particles in the atmosphere, small enough to be inhaled, pose possible health risks. Ozone formed in the atmosphere from nitrogen oxides and hydrocarbons is toxic to plant life and harmful to human health.

Secondary air pollutants have several factors in common: (1) they can form over periods of hours to days and travel hundreds to possibly thousands of kilometers; (2) they cannot be controlled directly, but only by controlling the primary pollutants from which they are formed (or intermediaries that determine their rate and transformation); (3) different secondary pollutants result from the same primary pollutants (e.g., nitrogen oxides can react to form both nitrates and ozone); and (4) they manifest themselves in several ways (e.g., sulfate contributes to both acid rain and visibility reductions).

Risk and Uncertainty

The concepts of uncertainty and risk recur throughout the public debate about controlling transported air pollutants. Each term is used in a variety of ways, both quantitatively and qualitatively, further complicating the debate. Their common usages—the concepts referred to in congressional hearings and the popular press—are examined here as a preface to outlining key uncertainties about transported air pollutants. Note that these usages often differ from conventions used within the decision theory and risk-assessment literature.

One of the major uncertainties, for example, is whether a reduction in sulfur dioxide emissions will lead to a comparable reduction in acid deposition. Using available data and theory, scientists devise mathematical models to attempt to answer this question. Uncertainty, in a quantitative sense, is a measure of how far wrong these models might be. Probability and statistics are used to quantify the uncertainty, providing numerical estimates of the likelihood that the outcome will fall within a specified range.

When the likelihood cannot be expressed numerically, professionals must rely on their subjective judgment. In these cases, which stem from limitations in data or understanding, qualitative terms such as "potential," "plausible," and "credible" are used to describe the probability that such an outcome will occur.

Risk is a broader concept than uncertainty. While uncertainty describes the likelihood of some particular outcome(s) occurring, risk combines the likelihood that each possible outcome will occur with the consequences of each outcome. Risk is difficult to conceptualize because the word is often used solely to describe consequences, as in "what is at risk if a flood occurs." This usage, however, refers to only part of the concept. Risk to a society is a combination of what is at risk—e.g., what is at stake if a small, medium, or large flood occurs—and the uncertainty (or probability) that each of these floods will occur.

Though risk can be quantified in many ways—and experts disagree over which way is best—all methods combine a series of alternative outcomes (e.g., the damage done by different size floods) with estimates of the probability of each outcome (e.g., flood) occurring. As might be expected, quantitative measurements of risk require quantitative estimates of uncertainty.

But for transported air pollutants, knowledge is currently insufficient to delineate the exact range of uncertainties. Though we can estimate, for example, the number of lakes and streams currently affected, we do not have enough data to know how much in error these estimates might be. Moreover, when we try to predict future damage, our abilities to distinguish among alternative, plausible scenarios are cruder still. Often researchers can only determine whether the answers are clearly wrong, rather than whether they are correct.

Key Scientific Uncertainties

Five key uncertainties are especially relevant to congressional decisions about transported air pollutants. These include controversies about (1) the extent and location of current damages, (2) future damages (whether they are cumulative and/or irreversible), (3) the geographic origins of observed levels of pollution, (4) the effectiveness of emissions reductions for reducing observed levels of transported pollutants, and (5) whether a research program will provide significant new results.

These scientific uncertainties affect several policy concerns, including: (1) making air pollution control policy as fair as possible, i.e., providing some legal recourse to those bearing the risks of damage, and (if a control program is adopted) distributing costs of control fairly; (2) intergenerational implications resulting from the risks of cumulative and irreversible damage; (3) weighing the risk of damage against gains that might be achieved by waiting for better information or improved technology; and (4) ensuring that the benefits, in the broadest sense, of a control program justify the cost.

1. Uncertainty About the Extent and Location of Current Damages

Scientists are certain that transported air pollutants have caused *some* damage. At issue is the extent of the damage, whether it is fairly localized or widespread, and which resources are affected. For example, there is little question that acid deposition damages lakes and ozone harms crops. The uncertainties revolve around how many lakes and streams and what quantity of crops. For these and other concerns, the risks of extensive damage over large parts of the United States are substantial. Later sections of this chapter present estimates of the risks in a number of affected areas for the eastern United States overall.

For certain other concerns, such as the extent of damage to forests from acid deposition, the uncertainties are so large that it is difficult to describe the patterns or magnitude of the risk. Damage to forests from ozone, and the effects of toxic metals released into drinking water due to acid deposition, also fall into this category. These types of risk are also discussed, but in very general terms, later in this paper.

2. Uncertainty About Future Damages

Growing scientific recognition that transported pollutants are linked to observed damages makes concern for future damage very real, even if current damages are not extensive. Of particular importance is the extent to which damages are cumulative and irreversible.

Cumulative Damages. For some resources, damage from transported air pollutants might worsen over time if pollution remains at about current levels. For short-lived species, such as crops harvested annually, this is not a relevant concern. Damage to this year's crop from ozone—and potentially from acid deposition—is caused only by current levels of pollution. For longer-lived species such as trees, however, cumulative damage is of concern. Potential ozone-induced damage to forest foliage and acid-deposition-related damage to forest soils may be small in any one year, but continued stress over many years may ultimately reduce productivity.

The portion of current aquatic resource damage attributable to the cumulative effects of deposition over many years is also uncertain. One major unknown is the degree to which years of exposure diminish the ability of soils in the surrounding watersheds to neutralize acid deposition.

The extent to which damages accumulate relates directly to the policy debate over the risks of delaying control action. If damages are not significantly cumulative, delaying control action while awaiting further information would not increase the level of damage from year to year (assuming that the level of pollution remains fairly constant). For cumulative effects of transported pollutants, however, the longer control actions are delayed, the greater the damage.

Irreversible Damages. A closely coupled concern is whether damage can easily be reversed, and if so, over what time period? Again, once ozone levels are reduced below the level at which significant damage occurs, direct crop damage is easily eliminated the following year. At the other extreme, potential damage to monuments or other unique art objects from acid deposition is irreversible.

If acid deposition is reduced, the loss of fish populations may be reversible over the time required to reestablish viable aquatic communities, either by planned restocking or by natural means. However, if surrounding soils have lost their acid-neutralizing capability, fish may not return for decades unless the water body is periodically treated with lime. Similarly, if acid deposition causes nutrients to be lost from a forest floor, forest productivity may be impaired for many decades. The effects of a severe series of ozone episodes, or the potential cumulative effects of acid deposition, will persist in the forest community until a new forest grows.

Scientific uncertainty about the reversibility of damages influences the policy debate in two ways. First, for reversible damages, one would ideally like information about the rate and extent of resource improvement in response to alternative levels of pollution reduction. The benefit of reducing ozone and acid deposition for crops and aquatic resources will be presented later in this paper, although considerable uncertainty surrounds these preliminary estimates.

Second, uncertainties about the irreversibility of resource damage create controversies about delaying control action until better information is obtained. If damages are both significantly cumulative and irreversible, waiting for better information would cause resources to be irretrievably lost for this generation and the next several to follow.

3. Uncertainty About the Origin of Observed Levels of Transported Pollutants

Pollutants Leading to Acid Deposition. The previous sections have outlined the uncertainties surrounding the broad question of potential benefits of reducing current levels of transported pollutants. If one wishes to reduce current levels of pollutants, the next logical point to raise is, "where are they coming from?" Three questions are important here: (1) whether the precursor pollutants are of natural or man-made origin; (2) whether they come from local sources or from distances exceeding hundreds of kilometers; and (3) whether it is possible to define the geographic origin of pollutants deposited in a particular region.

The first two questions are *not* key uncertainties. Though pollutants of natural origin cause some acid deposition, deposition over large areas of the eastern United States far exceeds the level attributable to natural sources. In addition, while local sources (i.e., within 30 miles) do contribute to acid deposition, most analyses indicate that a large share of the deposition originates from both medium-range and distant sources (i.e., more than 300 miles away) as well.

The key uncertainty is whether current understanding permits scientists to assess reliably how much of the deposition in any one region originates from emissions in any other. Computer models of varying sophistication are available to perform such analyses, but their accuracy in portraying this relationship is uncertain. The complexity and inherent variability of the atmosphere makes it unlikely that models will ever be able to predict how much one particular source of emissions contributes to deposition in a small area. However, these models can be used to assess current annual-average patterns of pollutant transport over large geographic regions. On this scale, models reproduce observed patterns of annual sulfur deposition in rainfall reasonably well. Their capacity to synthesize extensive meteorological data makes these models the best available tools for describing the current relationship between emissions in one area and deposition in another.

Model adequacy is a critical element in several suits and petitions by states to control transported air pollutants under the interstate provisions of the current Clean Air Act. EPA has adopted the position that available models are inadequate to address the question of long-range pollution transport; consequently, the agency does not currently consider the effects of pollution that travels beyond 50 kilometers (30 miles). However, the petitioning states assert that available long-range transport models reflect the state of the art and should therefore be used by EPA.

In addition, congressional attempts to focus reductions in pollutant deposition on identified, sensitive regions would rely to a large extent on model-based abilities to determine the sources of such deposition. However, no amount of accurate, detailed modeling information would be sufficient to eliminate policy disputes over which region's resources should be protected, or which region's emissions should be controlled.

Because many regions have sensitive resources, and all of them receive some deposition from each of many emitting regions, a variety of possible strategies could be used to reduce deposition on sensitive resources. For example, decision makers, in response to equity considerations, could develop control programs under which regions responsible for "large" shares of deposition in "several" sensitive regions bear the "greatest share" of emissions reductions. Whether the broad regional patterns of transport described by the models are accurate enough for this purpose is as much a policy question as a scientific one.

Pollutants Leading to Ozone Formation. The same three questions apply to ozone: whether the precursor pollutants are of natural or man-made origin, whether they are of local or distant origin, and whether it is possible to pinpoint the geographic source of deposition in a specified region. As with acid deposition, ozone levels over broad regions of the United States exceed natural background levels. Though locally produced ozone is a problem in many of the nation's urban areas, the chemical precursors of ozone are known to travel long distances, and elevated ozone levels are found in areas remote from sources of these precursors. However, models of long-range ozone transport are in a very rudimentary stage of development. Because the chemistry of ozone formation is very complex, major uncertainties exist over the geographic origin of elevated ozone concentrations in rural areas.

4. Uncertainty About the Effectiveness of Emissions Reductions for Reducing Observed Levels of Transported Pollutants

Uncertainty about the effectiveness of reducing emissions as a means of controlling transported pollutants stems from two major unknowns. The first—uncertainty about how well current models describe the relative contribution of one area's emissions to another's pollution—has been explained in the previous section. This source-receptor relationship cannot be defined precisely, but existing models can synthesize extensive meteorological information to provide broad regional descriptions. Its implications for policy are to create uncertainty about the extent to which emissions reductions in any one area will reduce pollution in another area.

The second unknown concerns the chemical processes that transform pollutants in the atmosphere. Numerous complex chemical reactions are involved in the formation of ozone, airborne sulfates, and acid deposition. Uncertainties exist about how effective reducing emissions of each precursor pollutant will be in controlling the transformed products.

For example, emissions of sulfur dioxide from fossil-fuel combustion are transformed to sulfate, a major constituent of acid deposition. Reducing the total amount of sulfur dioxide emitted will undoubtedly reduce the overall amount of deposited sulfate, but the reduction in deposition that will result from a given emissions cutback is difficult to quantify. In addition, scientists are uncertain that reducing sulfur dioxide emissions alone is the most efficient way to control acid deposition. Because other pollutants (such as hydrocarbons and nitrogen oxides) can enhance or impede chemical transformations of sulfur dioxide, simultaneous control of other pollutants might reduce acid deposition more efficiently.

Most analyses indicate that cutting back regional sulfur dioxide emissions alone will significantly but not quite equally reduce the amount of sulfur deposited in various forms. How much less the reduction in deposited sulfur will be, and whether this response can be increased by controlling emissions of other pollutants, is uncertain.

Uncertainties about which pollutants to control, as well as the relationship between emissions reductions and deposition reductions, complicate the policy objective of reaping the greatest possible benefit from the costs of controlling emissions. The inability to project deposition reductions precisely raises questions of whether the benefits of control will be realized to the extent, and in the locations, desired. Under these circumstances, a control program designed and implemented today may be more expensive than necessary.

5. Uncertainty About Whether a Research Program Will Provide Significant New Information

One of the most difficult decisions facing Congress is whether to act during this reauthorization of the Clean Air Act, or to wait for results from ongoing, multimillion-dollar federal and private research efforts on acid deposition. While the research efforts are intended to reduce the uncertainty discussed above, how much new insight five to ten years of further research will provide is unknown.

For example, years to decades are required to observe changes in many ecological processes. Patterns of crop yield and forest productivity typically vary from year to year; separating the effect of acid deposition from expected annual fluctuations requires many years of data, and normal year-to-year variations are expected. The processes of soil formation and depletion proceed on the time scale of decades. These mechanisms can be accelerated artificially and studied in laboratories, but the extent to which such experiments reflect real-world conditions is uncertain.

Complexities in both atmospheric meteorology and chemistry make research on atmospheric processes an uncertain, multiyear effort. Ongoing efforts to evaluate the current generation of computer models have already taken nearly two years. Efforts to develop new transport and transformation models are underway; the number of years needed to significantly improve existing models is unknown.

Uncertainties about the progress of research programs are an important factor when considering the timetable for policy decisions. Given the planning, contracts, construction, and so on necessary to achieve significant reductions in pollutant emissions, a decision to control emissions now may still require ten or more years to implement. Waiting five to ten years for the results of a research program before deciding to control emissions increases the time required to reduce deposition from ten years to 15 or 20. Such a delay increases the risk of resource damage but reduces the risk of inefficient control expenditures. Congress could avoid this delay by mandating controls now, while retaining the option to change the law if new research results show such action to be preferable. However, to achieve compliance within a decade, expenditures would have to begin within five to seven years. If results suggesting an alternative control strategy appear much after this time, major expenditures may be irrevocable.

The Risks of Damage and the Risks of Control

This section will present the remaining half of the concept of risk: what is at risk from controlling or not controlling transported air pollutants. As discussed previously, the risks from transported air pollutants are amenable only to qualitative descriptions. Certain aspects can be described numerically, for example, the numbers of lakes and streams currently altered by acid deposition, or the human-health effects of airborne sulfates and other small particles. But given the substantial uncertainties associated with available data and theory, even these numerical descriptions should not be viewed as exact, quantitative answers, but rather as qualitative estimates. Wherever possible, numerical estimates of the risks to a region's resources, health, industry, jobs, and consumer pocketbooks are presented. However, one should not be misled by the apparent precision of the numbers. The information presented is intended to convey approximate outcomes—in some cases with unknown margins of error—and broad regional patterns.

Different risks are associated with the two major congressional options: maintaining the status quo (no congressional action) or mandating further emissions reductions. The risks associated with each of these options will be discussed separately.

1. No Congressional Action

a. Resources at Risk. Two factors determine the potential for resource damage from transported air pollutants: (1) the amount of pollution to which the resource is exposed, and (2) the sensitivity of resources to acid deposition and ozone. Resource sensitivity varies from resource to resource and among the specific plants, animals, soils, and materials within each resource category. Some resources are directly affected by the pollution received. For example, crops are affected by exposure to ozone, and some building materials are affected by acid-producing substances deposited on their surfaces.

For other resources, characteristics of the local environment also determine sensitivity. Lakes and streams are affected not only by the acid deposition that falls directly on the surface of the water, but also by the amount of acid-producing substances that eventually travel through the watersheds. The acid-neutralizing capabilities of the soil and the bedrock in the watershed help determine the susceptibility of water bodies to acid deposition.

Aquatic Ecosystems. More is known about how acid deposition affects aquatic ecosystems — and the current extent of these effects — than for any other resource. Substantial evidence indicates that acid deposition alters water chemistry in sensitive lakes and streams. Most vulnerable are small lakes and streams in watersheds that have little capability to neutralize acid, due either to the chemical composition and/or thinness of the soil, or to terrain that is so steep or rocky that rainfall runs over it before acid can be neutralized.

Fish are sensitive both to the acidity of water and to toxic metals, primarily aluminum, that are released from the watershed under acid conditions. When waters become more acid than pH 5, most fish species are eliminated and major changes in lake ecosystem processes frequently occur. High acidity, toxic concentrations of aluminum, or some combination of the two appear to cause these changes.

Based on the predominant soil and geological characteristics in each county, OTA classifies about 25 percent of the land area of the eastern 31-state region as allowing the transport of acidity through a watershed to lakes and streams. Within these areas are approximately 17,000 lakes and 112,000 miles of streams. These figures provide an upper bound of the aquatic resources at risk. Only a portion of the total number of lakes and streams in the sensitive regions should currently be considered sensitive to acid inputs. Small lakes and stream segments are, in general, most susceptible. However, local variations in geology, soil conditions, and topography affect the land's ability to prevent acidification of water bodies. Though some areas have geologic characteristics that make them vulnerable to acidification, they may not receive enough acid deposition to alter the water quality of lakes and streams.

OTA used available water quality data from about 800 lakes and 400 streams scattered throughout eight states in the eastern half of the United States to estimate the number of lakes and streams currently sensitive to acid deposition. About 9,500 lakes and 60,000 miles of streams are currently considered to have water quality that, given sufficient acid deposition, might end up as acidified — a best guess encompassing about half the upper-bound estimate presented earlier.

OTA also used available water quality data, correcting for the percentage of lakes and streams that may be naturally acidic, to estimate the portion of these sensitive aquatic resources that have already become acidified, or have so little acid-neutralizing capability left that they are extremely sensitive to further acid deposition. About 3,000 lakes and 23,000 miles of streams fall in this category, about 20 percent of the lakes and streams found in identified sensitive areas.

It is not yet possible to estimate how many more lakes and streams could become acidified if current levels of acid deposition continue into the future. Two cumulative processes are important: (1) the potential loss of the soil's ability to neutralize acid inputs, and (2) for lakes, the accumulation of acidifying substances over periods of years to decades until lake water comes into equilibrium with the level of deposition received from the watershed. The latter process delays lake response to acid deposition, with greater time delays in larger lakes and lakes with low outflows.

Many of the currently acid-altered, smaller lakes and streams are likely to be in equilibrium with present deposition levels. By assuming that continued acid deposition will not degrade the soil's neutralizing capability—the best case situation—one can estimate water quality changes that might result from reduced acid deposition. OTA used a simple theoretical model to project how changes in acid deposition levels might change the numbers of acid-altered lakes and streams. If sulfate deposition were decreased by 20 percent, OTA estimates that 10 to 25 percent of these water bodies might show improvement. If sulfate deposition were reduced by 30 percent, about 15 to 40 percent might show improvement. If sulfate deposition were increased by 10 percent, OTA estimates that acid-alteration of lakes and streams might be intensified by about 5 to 15 percent.

Again, these are best-case estimates, assuming that the effects of acid deposition are not cumulative. If the effects are cumulative, some lakes and streams might still continue to acidify—though at a slower rate—despite the reduction levels discussed above.

Terrestrial Ecosystems. Both croplands and forested ecosystems are affected by a variety of transported air pollutants. The nitrates and sulfates in acid deposition may affect terrestrial ecosystems both harmfully and beneficially. Ozone, a gaseous pollutant toxic to plants, and its precursor pollutants (nitrogen oxides and hydrocarbons) can also be transported long distances into agricultural and forest lands.

Ozone may account for up to 90 percent of air-pollution-related damage to crops. How much crop damage is due to transported, rather than locally produced, ozone is uncertain. Observed effects include both damage to the quality of crops (such as leaf spotting) and reductions in crop yield.

Damage to crops from acid deposition under natural conditions has not yet been observed, although experiments using simulated acid rain have shown reduced yields and altered crop quality. Potential direct effects on vegetation are of greatest concern, since the qualities of chemical inputs used to increase agricultural yields can be adjusted to counteract soil acidification. The role of pollutant mixes may also be significant, since acid deposition seldom occurs in the absence of other pollutants. In some cases, it appears that the presence of sulfur or nitrogen oxides in the atmosphere makes crops more susceptible to ozone damage. However, other limited data suggest that pollutant mixtures may not be synergistic, or even that the presence of one pollutant may mitigate the effect of other pollutants.

To assess the risks to crops from transported air pollutants, OTA has estimated the benefits that might result if ozone were reduced to natural background levels. Similar estimates of the effects of acid deposition on crops, or of the effects of pollutant mixes, are not yet possible.

Data from field experiments were used to estimate ozone effects on crop productivity for peanuts (a sensitive crop), soybeans (sensitive/intermediate), wheat (intermediate), and corn (tolerant). These dose-response relationships were then combined with 1978 ozone monitoring data and 1978 agricultural statistics. Results suggest that if ozone levels had been reduced to natural, background levels in 1978, corn yields would have increased by 2 percent, wheat by 5 percent, soybeans by 13 percent, and peanuts by 24 percent. As measured by 1978 crop prices, this represents about a 6 to 7 percent loss of agricultural productivity, with almost two-thirds of the total impact on soybeans.

In forest ecosystems, transported pollutants could cause direct damage to trees, as well as affect the soils in which trees grow. Because trees are long-lived species, they are vulnerable to long-term chronic effects. Ozone damage to the foliage of many tree species has been observed, but the concentration of ozone necessary to cause damage not well known. About one-quarter to one-third of the forested land area in the eastern United States is exposed to ozone concentrations about twice the natural background levels.

Unlike agricultural soils, which are highly managed, forest soils can be affected both positively and negatively by acid deposition. The nitrates and sulfates from acid deposition can supply essential plant nutrients. In many areas, nitrogen in the soil is in short supply; on these soils nitrates in acid rain might improve forest productivity. Sulfur-deficient soils are quite rare.

Concern over the effects of acid deposition on forests stems from observed productivity declines and tree death in areas receiving elevated deposition levels. Pests, drought, and disease do not appear to be the sole cause of damage in such areas as the Adirondack Mountains in New York, the White Mountains in Vermont, the Pine Barrens in New Jersey, and forested areas of West Germany. Acid deposition is one of the possible factors under investigation, but has not been shown to be one of the causes of observed damage.

The major soil-mediated risks from acid deposition include: (1) in naturally acidic soils, mobilization of metals such as aluminum that are toxic to plants in sufficient quantity; and (2) the potential stripping of calcium, magnesium, and other nutrients essential for plant growth from the soil. Available information allowed OTA to identify those counties in which soils susceptible to these effects predominate; other locations in which these soil types occur to a lesser extent could not be determined.

Naturally acidic soils occur fairly extensively, being the predominant soil type in about half the counties east of the Rocky Mountains covered by forest or rangeland. The strong, freely moving acids from acid deposition can release aluminum and other toxic metals from these soils. However, whether toxic metals are released in sufficient quantities to affect forest productivity is unknown.

Acid deposition can remove essential nutrients such as calcium and magnesium from the soil, but the rate of removal—and the importance relative to other factors—is difficult to determine. Moderate pH soils are thought to lose nutrients at a faster rate than acid soils, but relatively few of these soils have low enough nutrient contents to be of concern in the near future. On nutrient-poor acid soils, especially those where sulfate can travel freely through the soil, further nutrient loss—even at a slow rate—from the soil, forest canopy, and decomposing plant material might be significant. About 20 percent of the eastern counties might meet these criteria, but whether this nutrient loss is significant enough to affect near-term forest productivity is unknown.

Data and knowledge limitations prevent a more detailed description of the risks of transported air pollutants to forests than given above. The extent of forested land either exposed to high ozone concentrations, or underlain with soils that may be sensitive to acid deposition, is the best available description of the forest resources at risk.

Other Resources. Air pollution damages a broad range of materials, including building stone, rubber, zinc, steel, and paint. Ozone is the pollutant that most affects rubber, while many other materials are chiefly affected by sulfur oxides. Humidity plays a key role in materials damage: dry-deposited sulfur dioxide and sulfate dissolve on moist surfaces (forming sulfuric acid) and may cause greater damage than relatively less acidic rain. Since air pollution is only one of many environmental factors (including temperature fluctuations, sunlight, salt, microorganisms) that cause materials damage, it is difficult to determine what proportion of damage it accounts for. Moreover, it is difficult to distinguish between materials damage caused by transported pollutants and damage caused by local pollution sources.

A number of analyses of the monetary costs of materials damage have made assumptions about the quality of sensitive materials exposed to elevated pollution levels and the effects of damage on replacement or repair rates. An alternative approach that does not rely on these assumptions was used in a recent study funded by EPA. The study used data on expenditures in 24 metropolitan areas and six manufacturing sectors to estimate the extra materials-related costs attributable to sulfur pollution. It concluded that reducing sulfur dioxide emissions in urban areas by about 25 to 30 percent below levels allowed by current primary standards would create total benefits of approximately $300 million annually for about one-half the households and about 5 to 10 percent of the producing sector in the United States. These results cannot be extrapolated to provide estimates of benefit to the nation overall, as the lack of necessary data precludes analyzing the remainder of the consuming and producing sectors in this manner.

Yet another resource at risk from transported air pollutants is visibility. Visibility levels depend on a number of factors, including humidity, man-made pollutant emissions, and such other factors as fog, dust, sea spray, volcanic emissions, and forest fires.

Pollutants impair visibility by scattering and absorbing light. Concentrations of fine particles—primarily sulfates and nitrates—can sufficiently interfere with the transmission of light to reduce visibility levels significantly. Sulfate concentrations correlate well with visibility impairment over large regions of the United States, especially during pollution episodes in the eastern United States. Nitrates rarely contribute substantially to visibility degradation in the East. However, in the western United States, windblown dust and nitrogen oxides can also reduce visibility significantly.

Trend analyses of visibility at eastern airports for the period 1950 to 1974 indicate that while wintertime visibilities improved in some northeastern locations, overall regional visibility declined. This trend is particularly pronounced for summertime visibilities. Summer, often the best season for visibility in the 1950s, is currently marked by the worst episodic regional haze conditions. Slight improvements in eastern U.S. visibility since the mid-1970s have recently been observed.

b. Risks of Health Effects. Extremely small particles, including airborne sulfates, are the component of transported air pollutants of greatest concern for human health. These small particles can travel long distances through the atmosphere and penetrate deeply into the lung if inhaled. Statistical (cross-sectional) studies show correlations between elevated mortality levels and elevated ambient sulfate levels. Whether sulfate is actually the causative agent, or merely an indicator of harmful agents (such as other particulates) found with sulfates, is unknown. To estimate damages caused by transported air pollution quantitatively, sulfate concentrations were used as an index of this "sulfate/particulate mix." In order to incorporate disagreements within the scientific community over the significance of sulfates to human health, OTA's analysis projected a range of mortality estimates for a given population exposure level. Using this procedure, OTA estimates that approximately 0 to 5 percent of the total deaths per year (with a point estimate of about 2 percent, or 50,000 deaths per year) in the United States and Canada might be attributable to current atmospheric sulfate/particulate pollution. This observed elevated mortality rate may reflect premature death caused by aggravation of preexisting respiratory or cardiac problems. If sulfur dioxide emissions levels remained the same through the year 2000, slightly higher levels of premature deaths might occur due to increases in population; a 30 percent decrease in emission levels by the year 2000 might reduce the percentage of deaths annually attributable to air pollution to 1.6 percent (40,000 persons). In each of these cases, estimates of the ranges of mortality extend from no deaths to about three times the point estimates reported above.

Researchers have not found consistent associations between health effects and outdoor concentrations of nitrogen oxides. High localized concentrations of nitrogen dioxide are considered greater cause for concern than ambient levels of transformed and transported nitrogen oxide pollutants. However, quantitative estimates of health-related damages caused by nitrogen oxides, or of the populations at risk from these pollutants, cannot yet be developed.

Acidified waters are capable of dissolving such metals as aluminum, copper, lead, and mercury, and releasing such toxic substances as asbestos, from pipes and conduits in drinking water distribution systems, as well as from soils and rocks in watersheds. No causal relationship has yet been demonstrated between acid deposition and degradation of drinking water quality although water samples in a few areas of New England and the Adirondacks receiving high levels of acidity have shown lead concentrations of up to 100 times health-based water quality standards. Mercury concentrations above public health standards have been found in fish from acidified lakes in Minnesota, Wisconsin, and New York. Elevated levels of aluminum and copper, two metals not considered toxic to the general public, have also been found in well water in the Adirondacks. Acidified municipal water supplies can be monitored and corrected quite easily, but acidified well water in rural areas is more difficult to detect and mitigate. Potential health effects from acidification of these water supplies and subsequent leaching of toxic substances remain of concern.

2. If Further Emissions Reductions Are Mandated

Of the major transported pollutants, congressional attention has focused on controlling acid deposition. Acid deposition results from both sulfur and nitrogen oxides; however, in the eastern United States, sulfur oxides currently contribute about twice as much acidity as nitrogen oxides. Several bills proposed during the 97th and 98th Congresses would require reductions in sulfur dioxide emissions of about 35 to 50 percent in the eastern United States (the 31 states bordering and east of the Mississippi River).

During 1980, sulfur dioxide emissions in the 31-state region were about 22 million tons. About 70 to 75 percent of the total—over 16 million tons—was emitted by electric utilities. About 10 to 15 percent of the total was emitted by nonutility combustion (primarily industrial boilers), with the remainder coming from industrial processes and other sources. Three states, Ohio, Pennsylvania, and Indiana, each emitted in excess of 2 million tons of sulfur dioxide per year, comprising 30 percent of the region's emissions. Six additional states, Illinois, Missouri, Kentucky, Florida, West Virginia, and Tennessee, emitted in excess of 1 million tons each, also comprising about 30 percent of the 31-state total.

Most current legislative proposals would place the greatest burden of emissions reductions on the utility sector and the midwestern states. The costs imposed by a control program include: (1) increased electricity costs to consumers, (2) loss of coal production and resulting unemployment in regions where high-sulfur coal is mined, and (3) financial strain to certain vulnerable utilities and industrial corporations.

The risk imposed by a control program is that some of these costs might prove unnecessary. A control program designed ten years hence might achieve the same level of protection at lower costs than one designed today, or the level of required protection might be more accurately identified. This section first presents the potential costs of control, followed by a discussion of the risk that the control program might not be as efficient or cost-effective as desired.

a. Utility Control Costs. The cost of controlling sulfur dioxide emissions rises with increases in the total tonnage eliminated. As greater removal is sought, moreover, the cost of eliminating each additional ton of sulfur dioxide also increases. OTA estimates the costs of reducing sulfur dioxide emissions in the 31 eastern states, in 1982 dollars, to be about:

- $0.5 to $1 billion/year to eliminate about 4 to 5 million tons/year
- $1 to $2 billion/year to eliminate about 6 to 7 million tons/year
- $2 to $2.5 billion/year to eliminate about 8 million tons/year
- $2.5 to $3.5 billion/year to eliminate about 9 million tons/year

- $3 to $4 billion/year to eliminate about 10 million tons/year
- $4 to $5 billion/year to eliminate about 11 million tons/year.

The estimates above are based on the costs of controlling utility emissions and *do not* include costs to offset any *future emissions increases* from utilities and industry. These costs can also be presented as percent increases in average residential rates for electricity, on both a regional and state basis. For example, a 50 percent reduction in utility sulfur dioxide emissions (8 million tons per year, about 35 percent below *total* regional emissions) would increase *average* residential rates by about 2 to 3 percent. Rate increases would vary by state, of course—consumers in some states would pay no increases, while others would pay over twice the regional average. For specific utilities, costs may be somewhat higher; for example, several Indiana utilities have asserted that their residential rates might increase by as much as 50 percent.

The cost estimates presented above are based on control strategies that limit allowable rates of emissions—sulfur dioxide emitted per quantity of fuel burned—similar to several bills proposed during the 97th and 98th Congresses. The range of cost estimates presented above reflects the differences in cost associated with alternative methods of allocating emissions reductions within states.

The cost of removing each ton of sulfur dioxide is generally lower when the rate of emissions is high. Consequently, total regional control costs are lower when emissions cutbacks are allocated to states on the basis of their emissions rates than when equal reductions are required for all states. This method of allocating reductions concentrates a higher proportion of the burden on midwestern states that burn high-sulfur coal. However, many areas with lower rates of sulfur dioxide emissions are already paying higher costs for generating electricity with less air pollution.

b. Effects on the U.S. Coal Market. Switching from high- to low-sulfur coals is one of the major available options for achieving substantial emissions reductions. Consequently, emissions reductions designed to control acid deposition create the risk of significant coal-market shifts by increasing the demand for low-sulfur coals at the expense of high-sulfur coals. Risks of production and employment losses occur almost exclusively in the eastern United States, where coal reserves are primarily of high-sulfur content. The western coal-producing states, and parts of Kentucky and West Virginia, contain low-sulfur coal reserves and may therefore benefit from acid deposition controls. For the United States as a whole, regulations designed to control acid deposition would probably not affect total coal production.

The magnitude of the shift to low-sulfur coal depends on the relative cost advantage of fuel switching as opposed to removing sulfur dioxide by technological means (e.g., scrubbers). Each method has clear advantages for some situations, but for many plants, the cost difference is modest enough to make predictions of future preferences highly uncertain. The projections of future coal production and employment shifts presented below are based on current costs and might shift significantly with the availability of new control technologies or changes in coal prices.

For emissions reductions of 10 million tons and greater, 1990 levels of production in the high-sulfur coal areas of the Midwest (Illinois, Indiana, and western Kentucky) and northern Appalachia (Pennsylvania, Ohio, and northern West Virginia) might decline to about 10 to 20 percent below 1979 levels. Estimates of production declines are averaged over these regions and thus may be greater or less in some states and counties than others.

For the low-sulfur coal areas of central Appalachia (eastern Kentucky, southern West Virginia, Tennessee, and Virginia) and the western United States, acid rain control measures are projected to expand coal production beyond currently projected 1990 levels. This effect is more pronounced in central Appalachia than in the West, due to its proximity to eastern markets.

In general, employment changes would follow changes in production. A 10-million-ton emissions cutback is projected to reduce employment in high-sulfur coal-producing areas by between 20,000 and 30,000 jobs from projected 1990 levels. About 15,000 to 22,000 additional jobs would open up in eastern low-sulfur coal-producing areas, and an additional 5,000 to 7,000 jobs in the West. The risk of unemployment is most severe in Illinois, Ohio, northern West Virginia, and western Kentucky; for these areas coal-mining employment is projected to decline more than 10 percent below current levels.

These employment trends would be accompanied by proportionate changes in direct miner income. At the national level, coal-market shifts would have minimal monetary effects, as benefits to low-sulfur regions are projected to balance out losses to high-sulfur regions. This level of aggregation, however, obscures the regional distribution of coal-related economic costs. Direct income effects for a ten-million-ton emissions reduction are listed below:

Northern Appalachia: $250 to $350 million per year *loss*;
Central Appalachia: $400 to $550 million per year *gain*;
Midwest: $250 to $400 million per year *loss*;
West: $100 to $200 million per year *gain*.

Total economic impacts of coal-market shifts—reflecting, in addition, indirect employment and income effects on other economic activities—may be two to three times greater.

c. *Effects on Utility and Industrial Financial Health.* While the costs of further controlling utility sulfur emissions are ultimately passed on to electricity consumers, they can strain the resources of financially troubled utilities that must initially pay them. Additionally, controlling industrial process and boiler emissions imposes the risk of rendering sulfur dioxide-emitting industries in controlled regions less competitive than their less- or uncontrolled counterparts in other geographic areas.

On the basis of average 1980 utility bond ratings and returns on common shareholders' equity, utility sectors in eight states appear to be relatively vulnerable to the additional capital requirements that could result from further sulfur dioxide control: Arkansas, Connecticut, Florida, Maine, Michigan, Pennsylvania, Vermont, and Virginia. If cutbacks are allocated on the basis of emissions rates, three of these—Pennsylvania, Florida, and Michigan—are likely to be required to reduce emissions significantly. Reduction requirements are likely to be moderate for Florida and Michigan; moreover, both states are currently characterized by regulatory policies that allow increased fuel costs and/or most construction costs to be passed on to consumers relatively rapidly. However, in Pennsylvania, which has the highest average sulfur dioxide emissions rate of these eight states (2.5 lbs. per million Btu's in 1980; eleventh highest of the 31 eastern states), regulatory policies (1) do not currently allow utilities to pass on the bulk of construction costs to consumers until the facility is operating, and (2) permit fuel cost adjustments only once a year. Thus, on a state-average basis, utilities in Pennsylvania would appear to be at greatest risk from the imposition of stricter sulfur oxide controls, assuming that their regulatory and financial conditions did not change substantially by the time controls were implemented.

Major increases in electricity costs due to increased controls might also hurt utilities by reducing demand for electricity. However, OTA-projected increases in average electricity rates associated with stricter sulfur oxide controls are not high enough to affect demand for electrical power appreciably. Preliminary analyses of industrial process emissions suggest some potential for economic dislocation in the iron and steel and cement industries if their emissions were to be strictly controlled. Stricter emissions controls, if imposed in the southwestern United States, could also create hardships for the region's currently depressed copper smelting industry.

Notes

1. *Acid Rain and Transported Air Pollutants: Implications for Public Policy* (Washington, D.C.: U.S. Congress, Office of Technology Assessment, OTA-O-204, June 1984).

2. OTA staff contributing in-house work included: Robert M. Friedman, project director; Rosina M. Bierbaum, assistant project director; Patricia A. Catherwood; Stuart C. Diamond; George Hoberg, Jr.; and Valerie Ann Lee. Much of the information contained in this chapter

draws on work from the following OTA contractors: Daniel Bromley, University of Wisconsin, Madison; Duane Chapman, Cornell University; L. D. Hamilton, Brookhaven National Laboratory; Walter Heck, North Carolina State University; Dale Johnson, Oak Ridge National Laboratory; Orie Loucks, The Institute of Ecology; Brand L. Neimann, University of Illinois; Richard J. Olson, Oak Ridge National Laboratory; Ed Pechan, E. H. Pechan and Associates, Springfield, Virginia; B. J. Finlayson-Pitts, California State University; J. N. Pitts, University of California, Riverside; Perry Samson, University of Michigan; David Shriner, Oak Ridge National Laboratory; Jack Shannon, Argonne National Laboratory; John M. Skelly, Pennsylvania State University; Dan Violette, Energy and Resource Consultants, Boulder, Colorado; Michael P. Walsh, Arlington, Virginia; Gregory Wetstone, Environmental Law Institute, Washington, D.C.

PART TWO

The Meaning and Role of Consent

4

The Role of Consent in the Legitimation of Risky Activity

Samuel Scheffler

The topic I want to discuss is the role of consent in the legitimation of risky activity. Philosophers have had relatively little to say about this specific topic, but they have had a great deal to say about the moral importance of consent in other contexts and in general. For this reason among others, it seems natural to approach the question of the relation between risk and consent by way of some more general reflection on the importance of consent as a moral notion.

The first thing to notice is that, while few people regard consent as a morally unimportant notion, there is substantial difference of opinion about the nature of, and reasons for, whatever importance it may have. And, I believe, the differing conceptions of the moral importance of consent that have gained prominence tend to support different conceptions of the role of consent in legitimating risky activity. In order to make this clear, let me begin by sketching three different views of the moral significance of consent. These views are certainly not mutually exclusive, and they may not be jointly exhaustive. In other words, one could accept any combination of them, and there may be others. But the three views I shall sketch are all familiar and influential, and the fact that different people accept different combinations of them is itself of considerable importance for this discussion.

On the first account, the importance of consent is instrumental. The belief that consent has instrumental value is based in part on the observation that, in general, people are less willing to be treated in morally unacceptable ways themselves than they are to treat others unacceptably. To the extent that this observation is correct, consensual processes and consent requirements may constitute effective tools for deterring conduct that is deemed morally unacceptable for reasons which may have nothing further to do with consent per se. Moreover, in various social and institutional settings, consent mechanisms may have additional good effects: the reduction of alienation, hostility, and other destructive attitudes; the development of individuals' self-esteem and sense of personal responsibility; the promotion of feelings of identification, loyalty, and solidarity; and so on. Thus although, on this account, it may not be feasible or even desirable for individuals to have the opportunity to consent to every single course of action that will significantly affect their lives, there are nevertheless some very good reasons for taking consent seriously.

On the second account, consent has intrinsic moral value because it is an important component of a good life. Although what counts as a good life may vary from person to person, on this kind of view, the lives we regard as good are thought to have at least this much in common: they all include opportunities for the individual to exercise his capacities for judgment, decision, and commitment, and to exert some degree of control over the way in which he allocates his time and energy. A life that was utterly devoid of opportunities to make choices among alternative pursuits and courses of action would be a seriously impoverished life, on this view, and not just because one is more likely to end up with satisfying pursuits if one has some say in their selection. The process of selecting has an added value of its own. To be sure, it is not the only valuable thing in life, and in concrete situations it must often be balanced against other goods; but it is at the very least one good thing among others, on this account.

On the third view that I want to mention, the importance of consent derives from an influential picture of individual rights. According to this picture, individuals possess certain rights quite independently of the decisions or policies of any legal institution or governmental body, and these rights may be overridden only in the most extreme circumstances, if at all. Thus laws and other social restrictions on personal conduct must not, if they are to be legitimate, infringe individual rights. However, as proponents of this conception of rights are well aware, it is often socially desirable to have people's behavior limited in ways that are incompatible with the unrestricted exercise of individual rights. So it becomes very important, from the vantage point of this conception, to find some morally acceptable mechanism for reconciling individual rights with the need for a certain amount of social coordination and regulation of behavior. It is at this point, of course, that consent emerges as a crucial notion. A policy or course of action that would otherwise violate a person's rights may not do so if that person has consented to the policy or course of action. Thus, on this view, consent has a distinctive kind of significance: it provides an acceptable way, and perhaps the only acceptable way, of reconciling the moral rights of the individual with the need for social regulation.[1]

I have already said that these three conceptions of the moral importance of consent should not be thought of as mutually exclusive or jointly exhaustive. It should also be said that, in sketching these conceptions, I have ignored some genuine substantive differences, as well as important differences of nuance and emphasis, among the various different versions of each. For the purposes of this discussion, however, I think that the rather coarse-grained characterizations I have given will suffice, for they will enable us to appreciate some of the sources of two very different attitudes toward the role of consent in the legitimation of risky activity.

According to the first of these attitudes, there is a presumption against imposing serious risks on people without their consent, for there is a presumption that doing so will cause more harm than good. If, however, the overall (actual or expectable) results of imposing some particular risk without the risk bearer's consent would in fact be better than the overall (actual or expectable) results of other courses of action, then the imposition of that risk may well be regarded as permissible, despite the failure to obtain consent. This attitude receives support from the first two conceptions of the moral importance of consent that I have sketched. In other words, it is the attitude one is likely to have if one accepts either or both of the first two conceptions, but rejects the third. Remember that according to the first conception, the importance of consent lies in its effectiveness as an instrument for promoting good states of affairs and avoiding bad ones, and that according to the second conception, consent is intrinsically important because the process of voluntary choice is part of any good life. Now, both of these conceptions provide support for a presumption against imposing serious risks on people without their consent, for they both provide support for the presumption that imposing such risks will do more harm than good. For one thing, the imposition of serious risks on people whose consent has not been obtained diminishes the extent to which those people can exercise control over their own lives, and thus, according to the second conception, diminishes their ability to live good lives. (This problem will of course be exacerbated if those who bear the risks are not even informed about them.) Other considerations are suggested by the first conception. Since risk-imposing agents are, in general, unlikely to care as much about the interests of risk bearers as the risk bearers themselves do, there is a high probability that risk imposers who do not obtain the consent of risk bearers will overestimate the advantages of risk imposition and underestimate its cost to the risk bearers. The danger of this kind of miscalculation will be most acute when, as is often the case, the risk imposers themselves are among those who will benefit significantly from the risk imposition. From the perspective of the first conception, the relatively high probability of such miscalculation serves to support a presumption against imposing serious risks on people without their consent whether or not consent is also thought to have intrinsic moral value. And such a presumption receives additional support from those considerations, also emphasized by the first conception, that pertain to the efficacy of consensual processes and mechanisms as means of promoting healthy personal attitudes and social relations. To the extent that the imposition of serious risks on people without their consent, whether by an individual, a group, or a government, tends to promote alienation and hostility, and to undermine social solidarity and the sense of personal responsibility, there will be additional good reasons, from the vantage point of the first conception, to regard such risk imposition as morally problematic.

Thus the first and second conceptions of the moral importance of consent support a presumption against imposing serious risks on people who cannot be construed as having agreed to bear those risks. At the same time, however, they do not support an absolute prohibition, or even a nearly absolute prohibition, against such risk imposition. On the contrary, if the importance of consent is taken to be that it constitutes one of the good things in life and may help to promote some of the others, the rationale for insisting on consent will be considerably diminished, or even lost entirely, in cases where better overall states of affairs can be achieved by eschewing it than by requiring it. So the first and second conceptions also provide support for the other half of the attitude I've described: for the view that although there is indeed a presumption against imposing serious risks on people without their consent, the permissibility or impermissibility of imposing a particular risk of this kind can be determined, ultimately, only by weighing the costs and benefits of the imposition against the costs and benefits of alternative courses of actions.

This attitude toward risk and consent is liable to be misunderstood unless two points are emphasized. First, although the permissibility of imposing a particular risk without the risk bearer's consent is said to depend on the costs and benefits of such imposition relative to the costs and benefits of other possible courses of action, there is no requirement that the relevant costs and benefits must be construed in narrowly economic terms. On the contrary, the attitude I've described is compatible with highly pluralistic views of human values, and with measures of value which are not purely economic. Of course, it is much easier to say this than it is to produce a measure of costs and benefits which is not purely economic, which is sensitive to whatever genuine diversity of value there may be, and which provides a practical basis for social decision making. That, presumably, is one of the reasons why the tendency to rely on purely economic measures of costs and benefits is as great as it is. Nevertheless, as far as the attitude I've described is concerned, costs and benefits may include emotional and psychological states of individuals, social relations of specified kinds, decreases or increases in the quality of human lives, effects on future generations, ecological effects, and so on. And if purely economic measures of costs and benefits are not thought capable of reflecting all of these things accurately or adequately, then the task is to try to find some more acceptable measure. Failing that, even the use of rough and highly intuitive techniques for balancing different kinds of costs and benefits may be thought preferable to reliance on a purely economic measure. In any case, the point to bear in mind is that it is no part of the underlying attitude I have described that costs and benefits must be construed in exclusively economic terms.

Second, although this attitude requires that comparisons be made among the relative costs and benefits of different courses of action, there is no requirement that the principle of comparison must be purely aggregative. In other words, there is no assumption that one course of action is to be deemed superior to another if and only if the total net (actual or expectable) benefit of the first course of action is greater than that of the second. Instead, the principle of comparison may be distribution-sensitive: its rankings of sets of costs and benefits may be directly determined, at least in part, by the way in which each set is distributed among different individuals.[2]

Taken together, these two considerations may serve to remind us that the willingness to balance costs and benefits which is embodied in the attitude just described need not be a mark of crassness or insensitivity. Someone who maintains that it is legitimate to impose risks on people who have not agreed to bear those risks, provided the benefits of doing so are sufficiently great, can at the same time insist that the notion of "sufficiently great benefits" must be construed in a plausible way. In particular, such a person can perfectly well insist that the phrase must be interpreted in such a way as to capture our actual views about what things count as benefits and as burdens, as well as our views about the relative desirability of different distributions of goods. Despite this, there will be those who feel that this attitude fails to attach sufficient importance, or the right kind of importance, to consent. These people are likely to feel drawn to the second of the two attitudes toward risk and consent that I wish to discuss. While the attitude just described is supported by the first and second conceptions of the moral importance of consent sketched at the outset of the paper, this second attitude is, with one important qualification that I shall discuss, supported by the third conception.

According to the second attitude, there is more than just a presumption against imposing risks on people without their consent. Except perhaps in the most extreme emergencies, it is not regarded as permissible to impose risks (or at least certain kinds of risks) on people, whatever the advantages of doing so, unless they can legitimately be regarded as having consented to bear those risks. Explicit consent is not always required, of course; tacit or implicit agreement is often sufficient. Even some kind of hypothetical consent may be thought acceptable on occasion. But unless it is clearly correct to say that an individual has in some way consented to bear a particular risk (of one of the relevant kinds, perhaps), or that he would so consent under suitable conditions, then to impose that risk on him is to do something wrong. The only possible exceptions are cases in which imposing such a risk is the only way of avoiding some terrible catastrophe. To be sure, risky activity often does have very great benefits, and the consequences to society of prohibiting all risky activity would be disastrous. But what this suggests to those who share the attitude I am now describing is not that risky activity can be justified by its benefits alone, but rather that since risky activities are often beneficial, it is important to have strong and effective consent mechanisms in society so that agreement about which such activities to pursue can be arrived at legitimately and efficiently.

Just as the first attitude I described is clearly supported by the first and second conceptions of the moral importance of consent, this second attitude may appear to be straightforwardly supported by the third conception. After all, the second attitude as I've described it revolves around the belief that some or all kinds of risk-imposing activities are such that they may not legitimately be undertaken without the risk bearers' consent, even if they would be highly beneficial. And the third conception views consent as important because it is thought to be the primary acceptable mechanism for reconciling individual rights with behavior that may be socially desirable but that stands in prima facie violation of those rights. Despite this, the assertion that the third conception supports the second attitude needs to be qualified if it is to be accurate.

We can see why this is so if we take note of the following considerations. The two attitudes toward risk and consent that I have described have their analogues in two corresponding attitudes toward harm and consent. On the one hand, there will be those who feel, roughly, that there is a strong presumption against harming a person without his consent, but that it may nevertheless be permissible to inflict a harm that has not been consented to if the overall benefits of so doing would be greater than the overall benefits of alternative courses of action. On the other hand, there will be those who feel, roughly, that there are at least some kinds of harms that it is never or almost never legitimate to inflict on a person without his consent, whatever the benefits of doing so may be. The first of these attitudes is supported by the first two conceptions of the moral importance of consent, just as the first attitude toward risk and consent is. And the second attitude toward harm and consent is supported by the third conception, for that conception regards individuals as having rights not to be treated in certain specified ways unless they consent to such treatment, and prominent among the modes of treatment thus restricted are harmful ones.

To say this, however, is to begin to see why the second attitude toward risk and consent is not as straightforwardly supported by the third conception as the second attitude toward harm and consent is. For while traditional lists of individual rights do typically restrict harmful activity, they are generally silent on the question of risk. As the traditional accounts of rights are usually formulated, in other words, the extent to which risk-imposing behavior counts as violating individual rights is completely unclear.[3] Suppose one of Smith's activities exposes Jones to a probability p ($p<1$) of a harm, and that harm would constitute a violation of Jones's rights if it were the inevitable result of Smith's actions. Does Smith's risky activity violate Jones's rights if, as it turns out, Jones is in fact harmed? Does it violate Jones's rights even if Jones is not harmed? How great must p be before the answer to these questions becomes "yes"? Does that depend on the magnitude of the harm? As one leading contemporary defender of rights has observed, "No natural-law theory has yet specified a precise line delimiting people's natural rights in risky situations."[4] Thus the second attitude toward risk and consent will be supported by the view of consent as a mechanism for legitimating behavior that would otherwise violate individual rights only if traditional accounts of rights are modified to include some or all forms of risk within the scope of the moral protection that the rights are meant to afford.

In sketching each of these two attitudes, I have once again provided rather coarse-grained characterizations, ignoring some important details and qualifications, and abstracting from certain variations of both principle and nuance. I hope that my sketches will nevertheless serve to provide a clear enough picture of the two basic attitudes, and of the ways in which they diverge, to allow for some discussion of their merits. Before initiating that discussion, however, it does seem to me important to make one further point of clarification. It should not be thought that those who share the first attitude are necessarily unsympathetic to talk about rights in any form. On the contrary, those who take that attitude may be willing to grant that rights have a legitimate role to play in moral thought, and may agree that the social, legal, and economic institutions of a just society will appropriately assign what are termed "rights" to the citizens of that society. They will differ from those who take the second attitude, however, in their view of the justification of rights, and in the extent to which they are willing to allow considerations of consequences to determine both what rights do and don't allow and when they may legitimately be overridden. In other words, those who take the first attitude will tend to see rights themselves, whether moral, legal, political, or economic, as justified by their consequences: by the interests their recognition serves to advance. As a result, they will typically assign consequences a much larger role in determining what rights do and don't allow, and in determining when rights may be overridden, than those who take the second attitude do.

Of the two attitudes I have described, I find myself more drawn to the first. For while the two conceptions of the moral importance of consent that support that attitude seem to me compelling, the third conception seems to me troubling, for a number of reasons. The first of these reasons concerns the picture of individual rights presupposed by that conception. In other words, it concerns the general view of rights relied on by that conception, and not the relation between rights so construed and consent or risk. Since that is so, and since space is limited, I will mention this first reason only briefly and not try to elaborate it in any detail. I will simply assert, though I have elsewhere argued, that it is very difficult to provide a coherent rationale for rights that is not itself consequentialist in nature.[5] But if the rationale for rights *is* consequentialist in nature, I do not believe that it will support the very strong kinds of claims about rights which are typically advanced by their defenders, and which would be necessary in order to sustain anything like the third conception. It will not, for example, support the claim that it is ordinarily impermissible to violate a right even in circumstances where doing so would prevent more numerous or more serious violations of the very same right, or other events at least as objectionable, and would have no other morally relevant consequences. Yet it is just this kind of seemingly paradoxical claim that is typically advanced by those who defend nonconsequentialist accounts of rights, and that would indeed have to be correct if the third conception were to be vindicated.[6] Thus, in the absence of a plausible nonconsequentialist rationale for rights, moral views that presuppose a nonconsequentialist construal of rights, as the third conception does, seem to me problematic.

My next reason for skepticism about the second attitude concerns the ability of rights theories like those presupposed by the third conception to treat questions of risk plausibly. I said earlier that the second attitude would be supported by the third conception only if traditional accounts of rights were modified to include some or all forms of risk within the scope of the moral protection that rights are meant to afford. What I am saying now is that it is not altogether clear how such modifications are plausibly to be made. A very great proportion of what we consider normal, everyday activities expose people to some risk of harm, even of quite serious harm. Driving one's car, playing softball, walking one's dog, having children, boating, riding one's bicycle, lighting a match, giving a gift—all of these are activities which, at the very least, often expose people other than the agents to some degree of risk. And virtually any type of activity can sometimes expose someone other than the agent to a degree of risk. Thus if it is maintained that people have the right not to have any risks imposed on them without their consent, the effect will be to rule out a good deal of ordinary life, since in most cases, as it seems to me, it simply will not be feasible to obtain the explicit or implied consent of all of those one places at risk by undertaking some ordinary risky activity.

If a rights theory is to be capable of handling cases of risk in a plausible manner, therefore, it must find some way of limiting the kinds of risk imposition that are said to violate individual rights. Two restrictions which sound plausible come readily to mind.[7] First, it should be noted that, in standard versions of such theories, not all cases of harming are thought to violate individual rights. If someone tries to kill me and I fight back, I am not ordinarily thought to have violated my attacker's rights even if I succeed in harming him. To take a somewhat different example, Jodie Foster may for all I know have harmed John Hinckley by spurning his overtures, but even if she did, on nobody's view did she violate his rights. Thus one natural restriction for a rights theorist to propose is this: if inflicting a harm of a certain kind on a person would not violate that person's rights, then imposing a risk of such a harm does not violate the person's rights either. A second natural restriction is this: even if inflicting a harm of a certain kind on someone would violate his rights, it would not violate his rights to impose a trivial risk of such a harm, provided that the harm does not actually come to pass.[8] It might be thought that, taken together, these two restrictions enable a rights view to deal plausibly with cases of risk. A person has the right, it might be said, not to have imposed on him nontrivial risks of harms whose infliction would violate his rights, even if the harms do not actually come to pass.

One difficulty with this formulation is that it depends on a prior set of criteria for determining exactly which forms of harmful activity violate rights and which do not. But although there are some categories of harmful activity which clearly are taken to violate individual rights (thrill killings, for example), and others which clearly are not (harms inflicted in self-defense, for example), it is not easy to find, in the literature of rights, any comprehensive set of criteria for drawing the distinction in the full range of relevant cases. And of course we really need to have such criteria before us if we are to assess a candidate risk principle which depends on them, as the principle formulated above does. The effect of that principle would be greatly to increase the kinds of activity prohibited by a rights theory; and unless we know which specific forms of activity the principle would serve to rule out, it is hard to know whether it represents a plausible way of treating questions of risk. But we cannot know which forms of activity it would serve to rule out until we know exactly which kinds of harming are viewed as violating individual rights and which are not.

What does seem clear is that very many ordinary activities impose risks that it would be quite misleading to call trivial, of harms that may be very serious indeed. (An example: Jones has a bad case of the flu. When she breathes, the flu virus gets circulated through the ventilation system of the apartment building where she lives. Among the other inhabitants of the building are a number of elderly people in rather feeble health. By staying in her apartment and breathing, Jones imposes a nontrivial risk of death on them, for the mortality rate among elderly people in feeble health who contract this kind of flu is, we may suppose, far from negligible. Another example: The Smiths decide to go away for the Labor Day weekend. They plan to travel by car and to take along their young daughter. By doing so, they will be imposing a small but, as it seems to me, a clearly nontrivial risk of injury or death on their daughter. Examples could easily be multiplied.) If that is correct, and if the rights theorist does indeed rely on a risk principle like the one under discussion, it means that the need for a comprehensive set of criteria for distinguishing those harmful activities that do violate rights from those that do not may be more urgent than has been generally realized. For such criteria will be needed by the rights theorist, not simply for the purpose of making relatively fine discriminations within the restricted category of directly harmful behavior, but also in order to assess the moral acceptability of an extraordinarily wide range of mundane but more or less risky activities. I am in no position to claim that such criteria cannot be provided, or that if provided they are sure to restrict risky activity in ways that sound implausible; but I do think that it is difficult to feel entirely comfortable about a rights-based approach to risk so long as we are lacking a clear statement of those criteria.

My final reason for doubt about the third conception of the moral significance of consent, and hence about the second attitude toward risk and consent, concerns the ability of the concept of consent to bear the weight placed on it by the third conception. What I mean is this. Although, according to the third conception, one's right not to be treated in a certain way may be surrendered if one consents to such treatment, it is also typically maintained that there are some conditions under which consent does not have this kind of effect. In other words, consent offered in certain kinds of contexts is regarded as morally invalid. Now, views differ widely as to the precise sorts of conditions under which consent is invalid. On one traditional view, only severe physical duress of certain narrowly specified kinds is thought to undermine consent. Thus, for example, if I agree at gunpoint to give away the contents of my wallet, my rights are indeed thought to have been violated, despite my consent. But, at the same time, the market behavior of individuals is treated, within quite a wide range, as morally valid consenting behavior, despite the fact that shares of power and wealth may be distributed in radically unequal fashion among the participants in the market, with the result that the range of choices genuinely available to some people will be much more severely constrained than the range of choices available to others. Many people have felt that this sort of view is implausible, and that if institutional arrangements constrain the choices of some individuals in certain fairly drastic ways, it becomes inappropriate to regard the choices eventually made by those individuals as morally decisive, for reasons not so very different from the reasons for which it is inappropriate to regard the muggee's agreement to hand over his wallet to the mugger as morally decisive. At the very least, many have felt, it becomes disingenuous to insist that the justification for accepting the outcomes of market processes lies exclusively in the consensual nature of participation in those processes.

Those who accept this criticism, but who also wish to remain within the framework of the third conception, are thus driven to articulate an alternative and more stringent set of requirements for morally valid consent. And these requirements notably include certain conditions that must be satisfied by those background institutions that help to shape the contexts within which consensual arrangements are arrived at. But while I am sympathetic to the motivation for moving away from the more traditional conception of valid consent, the idea of doing so while remaining within the framework of the third conception seems to me to have problems of its own. The main difficulty is that, as the alternative view's conditions on institutions get spelled out, it becomes evident that those institutions will also subject the choices of individuals to some quite significant constraints, albeit different constraints from those which the more traditional conception is willing to tolerate, and that the alternative view will also be willing to regard consent rendered subject to its preferred constraints as morally valid. Once this is recognized, however, it seems to me again disingenuous to retain the claim that consent per se is the central mechanism for reconciling the rights of the individual with the need for social regulation. And it seems to me much more straightforward and plausible to make the frankly evaluative claim that (the opportunity to) consent is a potentially important human good, but one whose value varies from context to context, depending in large part on the nature of the background institutions and constraints within which it is embedded. To say this, however, is to say that consent cannot serve as the sole vehicle for the justification of social institutions and patterns of social coordination, for its own value depends in part on those very institutions and patterns. And to say this, of course, is to abandon the third conception altogether.

For the various reasons I have outlined, I find myself uneasy about the second attitude toward risk and consent. And, on the whole, I feel more comfortable with the first attitude. There are, however, a number of qualifications and comments I would like to add. First, I take seriously the remarks I made earlier about the compatibility of the first attitude with measures of costs and benefits that are not purely economic, and with principles for comparing overall patterns of costs and benefits that are not purely aggregative. If the first attitude is indeed to be plausible, in my view, it must incorporate features of these two kinds. Second, the first attitude, as I have described it, does not distinguish questions about the legitimacy of risks imposed by different kinds of agents, e.g., individuals, corporations, governments, etc. These distinctions seem to me to be interesting and important ones, and a full discussion of them might well lead to certain modifications of the first attitude.

Third, there is one advantage of the first attitude which seems to me worth mentioning if only because it may appear somewhat paradoxical. As noted earlier, those who accept the third conception of the moral importance of consent, most notably those working within the social contract tradition in political philosophy, have emphasized the distinctions among explicit, tacit, and hypothetical consent. Sometimes, however, the interest of these varieties of consent has been to some extent obscured, because the very tradition that has called attention to them has also on occasion tended to distort them. Convinced that consent is the only legitimate basis for political obligation, and wishing to establish that most people in normal circumstances do indeed have political obligations, contract theorists have tended to see tacit or other forms of consent in all kinds of unlikely places, and one result of this, I think, has been some cheapening or devaluation of these notions.[9] Once one ceases to be committed to finding consent everywhere, one becomes, if anything, freer to appreciate it, in all its varieties and degrees, where it really can be found. Accordingly, if one ceases to regard consent as the indispensable means for avoiding rights violations, and views it instead as one highly effective instrument for making morally acceptable decisions about, say, risk, one should be more than willing to accept and experiment with a wide range of consensual and partly consensual devices. From the perspective of someone who has the first attitude toward risk and consent there is, for example, every reason to investigate the circumstances under which it really makes sense to speak of someone as having tacitly consented to bear a particular risk.[10] And there is also good reason to investigate the feasibility of certain kinds of quasi-consensual devices, such as the establishment, in certain situations, of what might be termed "risk-distribution panels": panels composed of individuals linked through personal interest to the various groups affected by a particular risk decision, and charged with the responsibility for helping to arrive at such decisions. The essential point is that those who take the first attitude need be no less willing, and may even on occasion be more willing, than those who take the second attitude to experiment with the full range of types and degrees of consensual arrangements.

Finally, in assessing the costs and benefits of risky courses of action, as recommended by the first attitude, we need to take our actual human reactions to risk very seriously and to beware of relying on technically attractive but humanly unrecognizable models of rationality. For example, two different hazards may expose people to risks which are numerically equivalent but which are not regarded as equally alarming, and rather than assuming, when such a case arises, that the differential perception is irrational, we should be prepared to examine the salient features of each hazard—the features responsible for the differential perception—and to entertain the possibility that the moral costs associated with the two risks may be different despite their numerical equivalence.[11] More generally, we should recognize that, just as theoretical models of rationality may teach us something about our attitudes toward risk, so too we may learn something about rationality by studying our reactions to risks of various kinds. No theoretical model of rationality, however persuasive in the abstract, automatically takes precedence over conflicting, but deeply entrenched, human attitudes. At the same time, of course, we should not regard our actual attitudes uncritically. Our attitudes toward risks of various kinds are influenced by a great many factors, and there will certainly be cases where reflective examination of such attitudes reveals them to be problematic in some respect. We need to remember, for example, that powerful social institutions and interest groups often have a significant stake in getting us to regard certain risks as "normal," and other risks as excessive and alarming. There is also a natural tendency, often exploited by one interest group or another, to regard familiar risks as normal and unfamiliar risks as alarming. It is even the case, as has been widely noted, that the same people may react differently to numerically equivalent estimates of a single particular risk, depending on which of the numerically equivalent formulations they are presented with.[12] For example, a one in five chance of contracting a certain disease might be deemed unacceptable, but a four to one chance of avoiding it judged acceptable. Considerations such as these should serve to discourage us from any uncritical commitment to our actual attitudes. In suggesting that we need to bring just the right mixture of sensitivity and skepticism to the assessment of our actual attitudes, I am not suggesting that this goal is easily achieved: only that this difficult goal is the one we should be aiming for.

Notes

1. Some who defend rights believe that it may on occasion be permissible to violate them provided the victim is compensated. I will ignore this complication in this essay, but I hope that this does not undermine my central points.

2. I discuss the notion of a distribution-sensitive principle at greater length in *The Rejection of Consequentialism* (Oxford: Clarendon Press, 1982).

3. Of course, some actions harm certain people and, by so doing, also expose the very same people to a risk of further harm in the future. The initial harm may be taken to constitute a rights violation and hence be prohibited, resulting ipso facto in a prohibition against that particular risk imposition as well. In this way, a rights theory may generate an indirect prohibition against the imposition of certain kinds of risks.

4. Robert Nozick, *Anarchy, State, and Utopia* (New York: Basic Books, 1974), p. 75.

5. In *The Rejection of Consequentialism*, Chap. 4.

6. See, for example, Thomas Nagel, "Libertarianism Without Foundations," *Yale Law Review* 85 (1975): 144, and Nozick, *Anarchy, State, and Utopia*, pp. 28–29.

7. These two restrictions are similar to those proposed by Judith Thomson in "Imposing Risks," Chapter 6, this volume.

8. On some imaginable views, the final clause might be omitted. In other words, it might be said that where inflicting a harm of a certain kind on someone would violate his rights, it still would not violate his rights to impose a trivial risk of such a harm, even if the harm does, against the odds, come to pass. For, it might be said, one *inflicts* a harm on someone only if one causes that person to suffer the harm. And on some views of causation, event *A* did not cause event *B* if *B*'s occurrence after *A* was highly improbable. This shows how a metaphysical position may affect one's moral views.

9. For an early expression of the same complaint, see David Hume, "Of the Original Contract," in *Hume's Moral and Political Philosophy*, edited by Henry Aiken (New York: Hafner Press, 1948), pp. 356–72.

10. Incidentally, Judith Thomson's Unpleasant Way example suggests that even determining when someone has explicitly consented to bear some risk may not always be unproblematic. See "Imposing Risks."

11. All of this is argued very persuasively in Douglas MacLean's paper "Social Values and the Distribution of Risk," in *Values at Risk*, edited by Douglas MacLean, Maryland Studies in Public Philosophy (Totowa, N.J.: Rowman & Allanheld, 1985).

12. See, for example, Amos Tversky and Daniel Kahneman, "The Framing of Decisions and the Psychology of Choice," *Science* 211 (January 30, 1981): 453–58.

5
Locke, Stock, and Peril: Natural Property Rights, Pollution, and Risk

Peter Railton

Introduction

Lockean natural rights theories have long been associated with laissez-faire policies on the part of the government, in large measure because of the sanctity they accord to individual rights, especially private property rights. However, I will argue that if one attempts to apply such theories to moral questions about pollution, they present a different face, one set so firmly against laissez faire—or *laissez polluer*—as to countenance serious restriction of what Lockeans have traditionally taken to be the proper sphere of individual freedom.

Curiously, Lockean theories also face a challenge from the opposite direction. They may be inadequately restrictive concerning the imposition upon others of unwanted risks that do not eventuate in actual property damage. As we will see, this challenge should be especially troubling for those who hold that Lockeanism gives expression to the Kantian idea of respect for persons.

I will consider various ways in which one might attempt to modify classical Lockeanism to avoid these difficulties, but it will emerge that these modifications generally raise more problems than they solve for the Lockean and result in views that lack much of the intuitive appeal of more orthodox Lockeanism.

In short, I will argue that Lockeanism, classical or revisionist, may be incapable of striking an appropriate balance between restrictiveness and permissiveness in matters involving pollution and risk. This failure raises doubts about the adequacy of a Lockean framework to our moral universe.

A Lockean View[1]

A simple, appealing picture of morality informs much contemporary thought and action. On this view, individuals have certain natural rights which give them freedom to act in certain ways and oblige others not to interfere. The archetype of such a right is the right of private property. If Harlan has exclusive ownership of a pumpkin, it is his to do with as he pleases, and no one may rightfully take it from him or hinder him in his enjoyment of it. His right entitles him to exclude others from making any use of his pumpkin to which he does not consent. Of course, Harlan's property right is limited by similar rights of others. He cannot, without permission, rightfully lob his pumpkin onto another's porch. He is free to transfer his pumpkin to another, or to give him use of it, and although such contracts or gifts, once made, bring with them new obligations to carry out promises rendered, these further limits are self-imposed.

In the classic, Lockean form of this view, individuals have some property rights wholly independent of civil law: property in one's own body and its capacities, and a right to appropriate common property for one's own use by mixing one's own labor with it, so long as one does not waste and "enough and as good" is left in common for others.[2] These initial, "natural" property rights form the basis for whatever further property rights individuals may acquire by harvesting the fruits of nature or exchanging goods or labor with others. In the fullness of time, some individuals may acquire more extensive property entitlements than others and may transfer this wealth to whomever they please, but all retain an equal, natural right not to be harmed in person or (other) property.[3]

From this Lockean view emerges an image of moral space akin to a map at a registrar of deeds. Individual entitlements or rights determine a patchwork of boundaries within which people are free to live as they choose so long as they respect the boundaries of others.[4] To learn one's moral obligations one need only consult the map. Would a given act involve crossing another's boundary?[5] If so, it is prohibited; if not, permitted.[6] This Lockean view is often called antipaternalistic because it holds that individuals are entitled to final say over what happens within their own boundaries. It is also often called libertarian, since it is so centrally concerned with preserving a field of individual freedom of choice. It is not, however, equivalent to the view that we should maximize individual freedom of choice: that is an aggregative, social goal, foreign to the Lockean picture. If the choices individuals make within their boundaries, and the mutual arrangements they make across their boundaries, do not result in a social scheme with maximum individual freedom, that is perfectly acceptable. Individual entitlements and decisions define the limits within which any social goal or policy may legitimately be pursued, including the policy of promoting freedom.

This Lockean view is opposed to balancing as well as aggregation. If I violate a boundary but in the process bring about some valuable result, that does not count as offsetting the original harm. For example, if the industrious Smith were to seize the laggard Jones's land, he might produce much more food, lowering food prices locally and making possible an improvement in local diet. But this improvement in efficiency would not undo the original violation of Jones's property right, even if Jones himself were ultimately made better off as a result. For it is up to Jones what becomes of Jones's land, and Smith would have acted without Jones's leave. Moreover, whether a given act violates one's property right is not a matter of whether one experiences unpleasant consequences from it. If I take an old pair of socks from your wardrobe without asking, I have violated your property right in them even if you had little use for them or fail ever to notice the theft. Of course, the fact that the object stolen is of little value may lead you to refrain from bringing the law down upon me, or may lead the law to be lenient about punishment. But I cannot plead that I have done nothing wrong if I have contrived to take another's property without ill effects. In this, as well as in its opposition to aggregation and balancing, Lockeanism is anticonsequentialist.

As I have said, this is an appealing picture. Some have argued that its opposition to aggregation and balancing of consequences gives expression to the moral separateness and uniqueness of individuals, while its opposition to paternalism expresses the moral autonomy of individuals. Indeed, this view may be seen as an attempt to capture the Kantian idea that individuals are the ultimate bearers of moral value, and that we should always treat individuals as ends, not as means alone.[7] If the view at first seems callous because it emphasizes obligations not to interfere with one another rather than obligations to assist one another, it must be remembered that the complement of my not being under an obligation to help another is that the other is not under an obligation to help me: we both enjoy a realm of free choice, within which we are at liberty to devote as much or as little of ourselves or our resources to others as we choose.

A strong historical connection exists between this Lockean view and the free market of classical capitalist theory. Individuals command property, which they may exchange by mutual agreement, and the standard of fair exchange is simply that the trade receives their free consent (in the absence of fraud). What is produced and how, how the fruits of production are distributed, and similar matters are left up to individuals and to their particular decisions about work, consumption, and investment. Social principles of "just distribution," overall efficiency, or utility maximization are not to be imposed upon this process, although proponents of classical capitalism have argued that in a properly functioning market, the result of individual decisions will tend to be the efficient use of resources, the maximization of total wealth, a distribution of wealth that largely accords with marginal contribution to its production, and many other social goods besides.

This Lockean picture has, I think, impressed itself upon the consciousness of virtually all Americans, even those who would reject it. It is therefore important to ask what happens when a Lockean view confronts problems of pollution and risk.

Pollution and Boundary Crossing

Lockean natural rights theories ought to be unequivocal about the moral impermissibility of many pollution-caused injuries. If I spray my lettuce with an insecticide that drifts onto your property, where you breathe it and develop a nervous disorder, I have crossed a boundary wrongfully.[8] I may not have intended this result, and it may not even have been something I could have foreseen (I did not know the stuff was dangerous to humans, perhaps), but these facts do not alter the fundamental one: I violated a boundary without permission or provocation. Unintended or unforeseeable violations may deserve different punishment from intentional, foreseeable ones, but on a Lockean view we have an objective obligation not to cross a boundary, intentionally or otherwise.[9] If I take your Buick thinking it mine, I am not a thief, but my possession of it is wrongful, and it must be given back intact. If I should damage it in the process, I would be obliged to repair it (as I would not be obliged to repair damage that happened to occur to my own car). Arguably, I may also owe you something to compensate you for any inconvenience my illegitimate taking caused you (as I would not be obliged to compensate you for inconveniences I cause you by the legitimate exercise of my rights). Violations of rights may not always warrant punishment, but we cannot "read backwards" from the inappropriateness of punishment to the nonviolation of a right, any more than we can "read backwards" from a judge's suspension of a criminal sentence to the nonviolation of a law. Precisely because Lockean views are so clear about the wrongfulness of crossing boundaries unless permitted or provoked, they are (in theory at least) quite strict about pollution-caused injuries to persons or property, very much restricting the kinds of polluting activity that might legitimately go on.

For a polluting activity to be permissible, it would have to be shown to involve no wrongful boundary crossing, even of the slightest extent. There is no room in a Lockean view for regarding minor injuries inflicted across boundaries as morally permissible, since, as we saw, whether a boundary is crossed does not depend upon the magnitude of the effect, or the value of what was affected. Petty theft is still theft. Moreover, it is quite irrelevant that a pollution-caused injury may be temporary, for example, that one may recover from exposure to an airborne toxin and be good as new. Knife wounds, too, often mend nicely.

Nor should it matter whether the victim makes, or fails to make, a special effort to avoid a pollution-caused injury. The burden is plainly upon others not to act in such a way that one can escape harm from them while on one's own property or on common property only by making special efforts. If Gale throws a knife across my lawn, and it strikes me on the leg as I go about my business, it is no exculpation (on a Lockean view, at least) for him to claim that I could have escaped injury had I been wearing chain mail or had I earlier sold my property and moved out of his throwing range. A steel-mill owner cannot escape blame by saying that those who do not like his sulfur emissions are free to sell their homes and move elsewhere. His responsibility is to stay within his own boundaries; if his mill produces gases that corrode the lungs of those who own property in the vicinity, or who happen to be on nearby common property, he has done wrong. This is so even if the mill has been around longer than the current residents or passersby. Someone who voluntarily moves into a high-crime (or high-grime) neighborhood may have acted unwisely, but he has not laid down his rights, and those who invade his person or property violate these rights. We may have less sympathy for someone who does not take certain precautions to avoid wrongful harm at the hands of others and may feel less inclined to come to his assistance, but the duty to mind borders is in no way diminished by some people's incaution.

The question of when we can legitimately interpret an action (or inaction) as waiving a right is an entangling one, as is the related question of what it is we may interpret such an action as permitting. Does someone who knowingly and voluntarily accepts a risky job thereby give consent to whatever harm may befall him in the workplace? Presumably not; there is still the possibility that the employer acts negligently or maliciously. If I accept a position with someone well known for cheating his employees, it will still be wrong for him to do such a thing to me. On a Lockean view I *am* free to sign a contract laying down a number of my rights, that is, giving another permission to cross certain boundaries that would otherwise separate me from him. Moreover, some Lockean views permit my failure to object to the actions of another, or to quit his property when I am free to leave, to be interpreted as tacit consent to what is going on. Locke himself believed that by living in a country from which one is free to emigrate one gives tacit consent to its system of laws and governance—a claim Hume would later ridicule[10]—but he presumably would not say that living in a risky neighborhood gives tacit consent to the crimes that might befall one, or that crossing the street at a busy intersection rather than walking four blocks to a quieter one makes one fair game for motorists. Lockean tacit consent to the state does not make it legitimate for the state to violate my inalienable right to self-preservation or to appropriate my private property. These individual rights constrain what a state may legitimately do even when it is founded on the express consent of its people. (It is something else if the people also consent to give the state free access to their property.) Similarly, my rights in my person and property constrain what people may legitimately do to me even if I choose to live dangerously. (It is something else if I also declare my property to be anyone's for the taking.) Some Lockeans, including Locke, have argued that there are some natural rights that no apparent act of consent—tacit or express—could actually waive.[11]

Consent is a natural place to look for room within a Lockean scheme to provide greater freedom of action with regard to pollution, but I would like to postpone further consideration of this possibility in order to continue exploring the question of when, in cases where neither express nor tacit consent is present, a polluting activity constitutes a boundary crossing.

Dispositional Harms and Risk

Among the effects of pollution are not only certain manifest injuries—property loss, illness, disability, death—but also increases in the probability individuals will suffer such injuries. How restrictive Lockean views are with regard to pollution will depend upon whether this latter sort of effect is also counted as a boundary crossing.

A Lockean may urge a fundamental distinction among ways of increasing the probability individuals will suffer harm: I may change the probability you will suffer a manifest injury by having an actual, causal effect on your person or property, or alternatively, by doing something potentially harmful that could causally impinge upon you. In the latter case, you and your property may emerge from the encounter with heightened risk wholly untouched, in the same condition you both would have been had the risky activity not taken place. In the former case, some actual physical change has been wrought by me in your person or property—e.g., the side-stream smoke from my cigarette has clogged your alveoli, making you more likely to succumb to a respiratory infection. As it happens, you may in the end not contract any such infection, but you have suffered a physical change as the result of my actions that renders you less resilient, more vulnerable than before. This change may be difficult to detect and may make itself known in the population only by aggregate statistics. However, for many pollutants, such as tobacco smoke, statistics do not indicate the existence of a threshold of exposure below which there is no effect on the probability of infection. So we may suppose without being too unrealistic that each time I send some of my tobacco smoke your way you suffer some small physical change, not for the better. It seems to me that if exposure to a pollutant reduces your ability to resist infections, take vigorous exercise, perceive the environment (owing to impairment of the senses), and so on, then it has damaged your health, even if you do not in fact happen to contract an infection, seek vigorous exercise, or make fine perceptual discriminations. That is, health is a dispositional as well as manifest state, and if your capacities have been reduced by my polluting activities, then I have not merely raised the probability you will suffer harm, I have also harmed you. I have caused an actual, though perhaps not readily detectable, harm to you that has the additional (and sometimes more disturbing) effect of raising your probability of suffering further, more evident harm.

On the other hand, if your neighbor carries out an activity wholly on his own property, which raises the probability you will suffer harm—e.g., operating an unsafe miniature fission reactor—and yet no actual harm results (the reactor does not malfunction) it would seem that no boundary has been crossed.[12] Let us call cases of this sort, where there is no actual physical change produced in a person or his property by an activity that nonetheless raises the probability he will suffer wrongful harm, the imposition of pure risk. In the purest cases of pure risk, the person whose probability of injury has been increased is wholly unaware of this circumstance, so that his life proceeds exactly as it would have had the risk-imposing activity never occurred. For example: you do not know of your neighbor's reactor and suspect nothing unusual. Things get more complicated when we allow awareness (or other sorts of indirect effects) of the risk-imposing activity, but let's avoid complication for the moment.

Most of the cases of pollution that have awakened interest are cases in which some actual harms are caused within the exposed population in addition to any pure risk imposed.[13] If it is impermissible to cause actual harms, then these polluting activities are impermissible whether or not the imposition of pure risk is itself a harm. As a practical matter, then, what a Lockean should say about the permissibility of most polluting activities will not be much influenced by questions about pure risk. Moreover, his strictness about border crossings should suffice to rule out a much broader range of such activities than we currently prohibit. However, at least some polluting activities may result in nothing more than the imposition of pure risk, and some activities that are of concern with regard to air pollution are worrisome more because of their riskiness than because of the actual harm they are now causing, e.g., the generation of power by nuclear fission. Further, pollution aside, many of the things we do impose upon others pure risk but only infrequently lead to actual harm, and it is of interest to ask what a Lockean might say about such activities in general.

Common Property

Again, we need a distinction. The fellow who operates an unsafe nuclear reactor entirely on his own property seems to cross no boundaries. What of the fellow who introduces some toxic substances into the atmosphere which, as it happens, no one ever inhales, although some are at risk of doing so? He has crossed no boundaries of private property, perhaps, but in the Lockean scheme the atmosphere is property, too: common property. By rendering a portion of the atmosphere toxic, he has in effect appropriated it from the commons, making it impossible for anyone else to use it without injury. For vividness, imagine that the toxins take the form of a cloud that floats intact around the earth's atmosphere. Locke permits such appropriation only under special conditions: one must mix one's labor with the property taken from the commons, one must not use it wastefully, and one must leave "enough and as good" in common for others. Let us suppose that the act involved mixing his labor and was not simply wasteful: the toxic substance is an unavoidable by-product of a process he was using to make a living.[14] Does he leave enough and as good in common for others?

The bare fact that no one happens to breathe that part of the atmosphere he has appropriated suggests that he left enough to go around. To be sure of this, we would have to be sure that no one had to make special efforts to avoid breathing the spoiled air. Assume that we do know this. Did he leave the air "as good" as before? We do not know exactly what Locke meant by this phrase, but let us suppose it to mean that the amount remaining in common after the appropriation is of the same quality as that which was originally appropriated. Here we may find that our polluter has transgressed even though no one breathed his toxins. If, for example, the bit of air he appropriated was cleaner than the atmospheric average at the time, he has violated the Lockean condition. Or suppose there is a general worsening of the air. He has (let us say) taken some air of 1982 quality and removed it from the commons. By 1992, there may not be enough air of 1982 quality to go around. That would not be entirely our friend's doing, of course, but Locke's condition could be interpreted in such a way that this should not matter; at least part of the scarcity of 1982-quality air is his doing.

The polluter could complain that it is unrealistic to imagine his toxins floating around as a cloud of quasi-private property; surely they would simply dissipate into the atmosphere, and surely it would be an exaggeration to say that he has privately appropriated all of the air into which these pollutants make their way. It is, in fact, unclear what a Lockean should say about this sort of case,[15] but let us grant that his original utilization of common property need not have the effect of appropriating for him all of the atmosphere subsequently tinged by this pollution. Thus, it would not be illegitimate for others to make use of this tinged air. It might, however, be harmful for them to do so, especially if the pollutants involved have no threshold of zero effect. By his initial use of common property in 1982, he will have had the effect that air regarded as common property in 1992 is less good. If there is not enough 1982-quality air available in 1992, then his original utilization seems impermissible on Lockean grounds.

Now our polluter may protest that eventually his toxins will, in effect, disappear, leaving the commons as good as before. If by this he means that they will settle out of the atmosphere, he does not strengthen his case, for then they may leave 1992 common land less good than 1982 common land. They might even fall on private property, an outright border crossing at any level of effect. If instead he means that they will become harmless with time, that of course depends upon the nature of the pollutants; some pollutants become *worse* health threats after undergoing chemical change or combination in the atmosphere. Moreover, even if a breakdown to harmless substances does occur, it will take time, so that there may be a period during which enough and as good has not been left in common. Most likely, what the polluter has in mind is the rather old-fashioned view that nature is so vast that his particular effect upon it is negligible, of trifling consequence.

This view is old-fashioned in at least two ways. First, no one today can fail to be impressed with the finitude of that part of nature we actually inhabit. Second, we now know that small causes needn't always have small effects. "There is evidence that cancers start from single cells and it is believed that a single molecule may be enough to start a cancer."[16] Even if one's polluting activities emit no more than one part per billion of a carcinogen, at this concentration there would be trillions of potentially cancer-causing molecules in a room-sized volume of air. Modern medicine aside, the polluter needs to be reminded that Lockean views do not say that whether a border is wrongfully crossed depends upon the magnitude of the effect. Taking from the commons, even ever so slightly, is taking the property of others.

Still, suppose that nature really were boundless. And suppose, too, as Locke did, that the provision against wastage drops out once an imperishable medium of exchange has been introduced. Would it even then be trivial that the requirement of leaving enough and as good be met? Nature may be infinite without all portions of it being equally accessible or equally worth having. It would hardly count as leaving "enough and as good" for future generations if they have to go to the ends of the universe or great effort to obtain it. If someone were to appropriate a bit of handily located and readily used common property, like the earth's atmosphere, leaving others plenty of good atmosphere frozen on the surface of a planet circling Alpha-Centauri, he would have violated the Lockean conditions.

There is, however, a more interesting point to be made. Suppose nature to exceed in extent what we could ever actually appropriate, and even to be equally valuable and equally accessible in its parts. Might we then take from the commons at will? If your *private* property exceeded in extent what you would ever actually use, I still would not be entitled to take a portion of it even though you were left enough and as good afterwards. On the classical Lockean view, I would not be entitled to take part of your property even if I substituted for it something of equal value, unless I had your permission to do so. It emerges that Locke's justification of private appropriation from the commons rests upon an assumption that common property need not be accorded the same respect as private property, even that we need not accord others the same respect as (part) owners of common property that we owe them as owners (sometimes, part owners) of private property.[17]

What is the justification for this asymmetry? For Locke, the argument involves religious and practical considerations. He believes that God gave man the world in common in order that he might use it for his survival. He notes that as a practical matter, however, we cannot survive without appropriating from the commons: anything I take from nature and consume to sustain my own existence is for that reason no longer available to others. Further, as a practical matter, I cannot get the consent of all owners of common property—all mankind—before appropriating from it. So if we are to survive and flourish, which Locke believes to be both God's will and a law of reason, some nonconsensual way of legitimating private appropriation is needed.

Even if we leave God out of it, this is a plausible argument. But how can it be the basis for a Lockean property right that entitles an individual to exclude others from his private property even under those circumstances in which, by taking from it, they might enhance their chance of survival while still leaving him enough and as good? So long as there is enough and as good available to him either in common or in his remaining private holdings, or in the two together, the argument seems incapable of generating a right of exclusion. What of the fact that mixed in with his private holdings is his labor, something by nature belonging to him? Well, in his initial appropriation from the commons he, too, took something by nature belonging to other people.[18]

All along, Lockeans have taken common property—and our rights in it—less seriously than private property. If I besmirch part of your estate, this is a boundary crossing even if that besmirching never affects you materially. What I do reduces the capacities of your estate by effecting a physical change in it, and this, we say, is a wrongful harm even if you never attempt to use these capacities. For example, if I bespoil an out-of-the-way corner of your land and thereby lower its market value, I have wrongfully deprived you of property even if you never notice the spoilation or the loss in value. Why isn't a polluting act that reduces the capacities of common property (perhaps lowering its market value, too, were it to be sold), a violation of the property rights we all have in the commons, even if no one is ever materially affected by this act? Consistency would seem to demand that we put the two sorts of property on the same footing, at least in this regard.[19] But we have two choices of footing: we may promote common property to the status of private property, or we may demote private property to the status of common property.[20]

If we seize on the first alternative, then we must regard acts that introduce pollutants into common property without the permission of all mankind as boundary crossings, even if no individuals ever happen to have their own private holdings infringed. It would be irrelevant whether a polluter's effect were small, or whether enough and as good were left in common; just as it is irrelevant to whether I may rightfully take your private property that what I take is of little value, or that you have enough and as good private property left over. Most of the polluting acts we have heretofore called impositions of pure risk would become boundary crossings because they would involve violating property rights in the commons, and therefore would be morally impermissible harms. The result would be an extremely restrictive position on pollution. Indeed, there would not be much room for pollution left: one could befoul one's own, private nest, but only if nothing seeps over a border with common or private property. Even the idea of a private nest would become problematic, for appropriation of any property beyond one's mere self would require universal consent from mankind, including future generations. To make this first alternative workable, it would be necessary to develop a powerful doctrine of tacit or hypothetical consent. That may seem to be clutching at straws, given the difficulties of these notions, but they must be grasped at, for one cannot sensibly embrace the conclusion that there is no justified private appropriation from the commons. We must breathe, after all.

The second alternative, of demoting private property to the status of common property, fits better with the original Lockean argument and has less chokingly restrictive implications. (It should be kept in mind that we are talking here of natural property rights. There may be good pragmatic reasons for according different treatment to private and common property in civil law. On a Lockean view, however, civil law must respect natural rights. So unless citizens were to contract into some special arrangement, a civil code would have to accord property rights in the commons at least as much respect as they are due in a state of nature.) On this alternative, one may acquire private property from the commons by meeting the Lockean conditions—mixing one's labor, not wasting, leaving enough and as good for others—but this private property may in turn be appropriated by another if he mixes his labor with it, doesn't waste it, and leaves the original owner with enough and as good (in private or in readily accessible common property). Private property would no longer be inviolable, and this may make the second alternative unattractive to many. But if one believes in natural property rights, has doubts about tacit or hypothetical consent, and wants to be able to draw a breath without asking permission, this may be the best one can do.

Pollution of private or common property, so long as it leaves enough and as good remaining, may be permissible on this alternative, and so it is less restrictive about pollution than any version of Lockeanism thus far discussed. Even so, it would prohibit the imposition of pure risk in those cases where this involves lowering the quality of common property. That is a more restrictive policy on risk than many who think themselves Lockeans would accept, especially if we understand the criterion "as good" broadly, to include not only the capacity of common property to support life, but also its aesthetic qualities: the clarity of the air, the naturalness of the landscape, and so on. Locke's view was that man was given the earth to enjoy, not merely to subsist on. Certainly, if someone physically changes another's private property in a way that reduces its aesthetic value, this is ordinarily regarded as a harm. If common property deserves equal respect, then ruining the aesthetic qualities of common property would be permissible only if the same were permitted of private property; since private property includes our bodies, I doubt many would accept the notion that aesthetic damage may be ignored as without moral significance. Similar remarks apply to, for example, the economic value of common property. Thus, if my factory's smokestack emits a noxious substance, which as it happens no one actually breathes, I may still have crossed boundaries impermissibly by reducing the value or capacity of common property as a sustainer of life, a source of aesthetic enjoyment, or an economic asset.

The phrase "enough and as good" is sufficiently vague to leave it indeterminate just how much of a reduction in restrictiveness the second alternative would effect. The phrase may even be ambiguous: must what is left be as good in total as what was before, or (more weakly) must only that which is "enough" be as good as what was before? A very loose reading of the phrase—or of the other conditions of nonwastage and "mixing one's labor"—would leave private as well as public property quite open for use without the owner's consent.[21] However, one would be able to gain freedom in appropriating from common property only to the extent that one grants others similar liberties with one's own private property. This trade-off is only reasonable, for in both cases one is taking something owned by others.

Risk and Responsibility

Although advocates of Lockean natural rights theories have favored laissez-faire government and free-market solutions to a wide range of social problems, we have seen that such theories in fact furnish the basis for very tight governmental regulations on pollution. The reason is straightforward: the function of a Lockean state is to enforce property rights by prohibiting and policing unconsented-to boundary crossings, and pollution violates such rights. Since natural rights constrain civil rights, it would be impermissible for the state ever to permit crossings of natural boundaries for the sake of economic efficiency, social utility, or the like, unless all members of civil society consented to such an arrangement. On the classical view, it would be usurpation for the state to permit crossings of an individual's boundaries without his consent, even if this individual would be a net gainer in the end. Paternalism is simply an especially insidious form of usurpation.

We have seen that the restrictiveness of Lockean theories applies to dispositional as well as manifest harms, as long as the dispositional harm is due to actual physical effects of the polluting act. We have also seen that even when a polluting act merely increases the probability of manifest or dispositional harm, without having an actual physical effect on others, it may yet be impermissible on a Lockean view. Exactly which polluting acts are impermissible will depend upon how one resolves the asymmetry of private vs. common property, or whether one finds a way of salvaging the classical, asymmetrical view. But on all plausible readings of these various forms of Lockeanism, the state should be much more vigorous in prohibiting and policing pollution than is now the case. A call for a return to Lockean property rights as the foundation of social justice is a call for greater, rather than lesser, governmental restriction on polluting activities. What else should we expect from a view that erects absolute boundaries around individuals and their possessions and makes individuals sovereign within these boundaries?

Yet in what remains of this chapter, I will try to argue that Lockean views, even in their most restrictive forms, may in some ways not be restrictive enough. A plausible moral theory, by my lights, would be less restrictive overall, but more restrictive in certain areas, in particular, with respect to the imposition of pure risk. Earlier we considered a person who imposes risks but seems to cross no boundaries of either common or private property, the man who operates an unsafe nuclear reactor next door. As long as the reactor functions normally, he keeps within the boundaries separating his private domain from all others. On Lockean views, his should be a morally neutral act.[22]

The Lockean accepts the deontological notion that some acts—e.g., the violation of a natural right—are intrinsically wrong, even when they happen to have good consequences.[23] Let us accept this idea for now. If any act is intrinsically wrong, it would seem intrinsically wrong intentionally to raise the probability innocents will suffer harm. This is just a probabilistic form of the familiar principle that it is intrinsically wrong to bring deliberate harm to innocents. Is intent essential? If we hold the description of an act, A, constant, then it would be odd to say both (1) it is intrinsically wrong to do A with intent to do A, and (2) A itself is morally neutral. It seems more reasonable to say that there must be something wrong with A in the first place, which explains why intending to do A is intrinsically wrong. In the case where A is "acting so as to harm innocents," this is readily granted by most deontologists. Should it not also be granted when A is "acting so as to raise the probability of harm to innocents"? If this is so, then operating (what one does not realize is) an unsafe reactor in one's basement is not morally indifferent after all, for we would certainly want to say that doing so with intent to imperil others by operating an unsafe reactor is wrong. The contrary urge we have, to say that there is nothing wrong with operating an unsafe reactor so long as one is intending only to operate a safe reactor, is the urge to displace attention from the evaluation of acts to the evaluation of agents. If the agent has reason to believe the reactor is no threat, he is doing nothing contrary to his subjective duty: his duty relative to what he believes (or has reason to believe) is the case. But there is also the question whether what he is doing is something he would be obliged not to do if he had full knowledge of the facts, i.e., whether he is failing to do his objective duty. In practice, we often take subjective duty as the best approximation of objective duty, but recognize that the two may fail to coincide when we are mistaken about the facts. Is the man operating an unsafe reactor doing his objective duty? Of course not. One is objectively obliged not to maintain an unsafe condition that threatens innocents (other things equal), even if this is being done without evil intent.

Can we explain what is morally wrong with such an act in terms of the crossing of boundaries or the violation of property rights? What territorial right is violated if someone acts within his private domain in such a way as to increase the probability another will suffer wrongful harm, yet no such harm actually results?

It seems that the Lockean must recognize there to be something wrong with acts that raise the probability others will suffer wrongful harm. One motivation for the Lockean scheme draws upon the idea of respecting others and their rights. It certainly would raise a question about the extent to which I respect your rights if I thought it permissible to expose them to arbitrarily high increases in the probability they would be violated, so long as these probabilities did not chance to be realized. (In many cases, even if I deliberately set out to harm you I could do no more than to raise the probability you will be injured: I may take a shot at you, but my aim is imperfect.) To avoid entanglement in questions about intent, let us say in a hypothetical mode that if I were to know that an act of mine would increase the probability another would suffer wrongful harm, then, other things equal, respect for that person would be a reason for not performing it. Respect for others is not simply a matter of not happening to violate their rights, but of taking some care that my actions not happen to do so. If I treat your belongings carelessly when they are on loan to me, you may legitimately feel that I showed inadequate respect for your property even if no actual damage happened to occur. Moreover, it seems incompatible with the Kantian dictum of treating others as ends, not as means alone, to think there is nothing wrong in pursuing one's own interests even when this involves exposing others to arbitrarily high levels of risk, so long as no boundaries actually are crossed. "A miss is as good as a mile" seems too expedient an attitude to be consistent with respecting others as ends in themselves.

However, acceptance of this argument would precipitate an important change in the simplest version of the Lockean picture, for it would mean that the rights of others are not simply side constraints determining an arena within which I am free to go about my business as I please.[24] Rather, I am under other-regarding obligations even on my own turf. My freedom to swing my arm does not stop at your nose, but at some point where I begin to show inadequate respect for you by putting your nose at too much risk. Lockeans face something of a dilemma here. On the one hand, if they do not take risk into account except when it involves the crossing of a boundary of common or private property, then although they preserve the simple, territorial picture of morality that attracts many to Lockeanism, they will fail to take into account all that we mean when we talk of respect for others or their rights. On the other hand, if they take risk into account, then they will face some large difficulties: either they must redefine the moral boundaries to make them more restrictive, or they must admit that there is more to morality than staying within one's boundaries. This last admission would open the way for a more thorough rethinking of the Lockean picture.

Let us call an act that raises the probability another will suffer wrongful harm as a causal outcome of one's own behavior, where we leave it open whether this harmful outcome actually obtains, an *endangering* act. We may ask whether risk presents Lockeans with a genuine dilemma by considering several ways in which a Lockean might attempt to treat endangerment within the original spirit of his view.

Revisionist Lockeanism

Self-defense. For a certain range of cases, Locke himself has developed a doctrine for dealing with risk prior to actual border crossing, for he wrote that an individual who

> declare[s] by word or action, not a passionate and hasty but a sedate, settled design upon another's life, puts him[self] in a state of war with him against whom he has declared such an intention, and so has exposed his life to the other's power . . . it being reasonable and just I should have a right to destroy that which threatens me with destruction.[25]

Someone who deliberately puts me at risk has, then, violated the side constraints even before any actual injury is inflicted, and I would be entitled to use force or otherwise violate his territory if this were necessary to stop him. The motivation for such a principle is clear enough: were I to have to wait until actual injury has occurred, I would be defenseless against many serious harms. The classical Lockean, then, may claim to have a doctrine to deal with the imposition of risk, even when the risk remains pure.

However, we have already seen that it is plausible to say that endangerment morally ought not to happen, other things equal, even when it is not intended as such, and so does not involve any sort of settled design or declaration of war. One may endanger innocently if one's acts pose a threat to the rights of others but one is not at fault for failing to realize this; one may endanger negligently if one's failure to realize the threat one's acts pose is culpable in some way (e.g., is the result of a history of carelessness or inattention); and one may endanger deliberately if one actually intends that one's acts imperil others. (There are other categories as well.) Locke's doctrine applies only to the last case and so gives us no basis for constraining the innocent or negligent endangerer. Of course, we ordinarily judge the character of the deliberate endangerer more severely than that of the negligent endangerer, and we may have nothing bad to say about the character of the innocent endangerer. But in all cases we may judge the endangering act as one that, other things equal, ought not to be done (assuming that a nonendangering alternative is available to the agent). What is needed, then, is a development of Locke's theory to cover all sorts of endangerment.

This development must perform two tasks. First, it must answer such questions as when endangerment is wrong, what counteractions are justified against it, and so on. Second, it must tell us what it is about (say) innocent endangerment that makes it wrong, since such endangerment may not involve any actual boundary crossings and does not involve evil intent. We will consider two proposals, focusing initially on how they would accomplish the second task.

(1) If we cannot locate the wrongness of innocent endangerment in the psychology of the endangerer, we may yet look to the psychology of those endangered. If it is a harm to step on someone's toes, should it not be a harm to cause the often more severe and lasting discomfort that fear of harm may cause? Fear may be as debilitating as physical injury and may even bring about a number of physical disorders. Why draw boundary lines so as to include trivial physical damage and exclude grave mental damage? Is this any more than a fetishism of the tangible?

These strike me as important questions for the Lockean to ask himself, but Lockeans may be suspicious of the notion of psychological harm.[26] One may, after all, be concerned to distinguish "real harms," such as damage to property, from "imagined harms," such as the offense others might take at one's ideas or habits. If it were morally required that we avoid innocently causing certain psychological responses in others, it would be difficult to imagine what a Lockean system of boundaries might look like. Psychological effects flow across existing boundaries in a marvelous variety of ways, some of which depend much more heavily upon how others regard us than upon what we actually do. A natural right that no one else act in such a way as to cause one psychological distress would radically change the character of a Lockean scheme of things.

This is not of itself a conclusive argument against such a right. Moreover, one certainly cannot argue that damage to property is in general more troubling to individuals, or more a sign of disrespect for them, or more likely to involve treating them as a mere means, than the psychological distress they may suffer at the hands of others. If anything, the opposite seems true. It is therefore something of a mystery why psychological effects play so small a role in Lockean views. After all, such views usually do incorporate a prohibition against fraud, which essentially involves a psychological effect.

I suspect that a number of long-standing convictions are at work: ideas and feelings may be viewed as simply unreal in a way that land or limbs are not; individuals are thought to have more control over their mental states and how these are affected by the acts of others than they do over their physical states, so it is more likely that they could contrive to manufacture mental harms;[27] physical harm is more publicly observable than psychological harm, so its authenticity, origin, and extent are more reliably assessable. In some cases, however, psychological damage is real, nonmanufactured, and observable enough.

It seems implausible to claim that my natural rights are violated by an otherwise innocent act of another that causes me fear if the fear is an irrational reaction to the act or if the fear, while rational, has as its object no piece of potential wrongdoing on the part of the other—e.g., if you inform me by word or deed that my neighbor is coming after me with a hayfork, thereby exciting in me considerable rational fear of wrongdoing, but not on your part. (Is there anything intrinsically wrong with causing the anxiety that fear involves if the fear would be rational in the circumstances? Does it matter whether the object of fear is human wrongdoing or some natural calamity? Of course, it seems wrong to torment people with fear—rational or irrational—but the wrongness here could be laid to one's intention to disturb or to one's negligence in attending to how one's behavior affects others.)

Let us consider only the simplest case: Should a Lockean admit a natural right that others not act so as to cause one rational fear that one will suffer wrongful harm from them? Someone might argue that this right would do more to restrict our behavior than it would to enhance the quality of our lives or our freedom of action. But this is not a Lockean argument, for it uses aggregative, balancing considerations about consequences to test whether a right exists. A Lockean might try a different sort of argument, seemingly popular today: we cannot live in a no-risk society, so we must learn to live with a degree of rational fear rather than obsessively try to eliminate it from our lives. However, a Lockean presumably would reject the comparable argument that since we cannot live in a crime-free society, we must accept some criminal activity as permissible. Natural rights should not lose their hold upon us merely because we cannot eliminate all violations of them; nor is the impossibility of eliminating all violations a reason not to seek to minimize violation.

I know of no convincing argument that a Lockean, concerned that individuals be respected as ends, could use to refute the claim that we have a natural right that others not act in such a way as to cause us rational fear of suffering wrongful harm at their hands.[28] Yet even admitting such a right would not wholly solve the Lockean's problem of capturing the wrongness of endangerment. For endangerment seems wrong even when it arouses no fear in those at risk, e.g., when they simply are unaware of their peril. We might reformulate the right as a right that no one act in a way that *would* awaken rational fear of wrongful harm from him were his actions to be known and their possible consequences grasped. But is that all that is wrong with endangerment? Would it make any difference to the wrongness of my playing Russian roulette on my sleeping roommate that he is someone who constitutionally feels no fear? In such a case, is the wrongness attributable to the fact that someone else, less fearless, would feel fear if he knew *he* were being exposed to such endangerment? Kantians, at least, would presumably deny this. They have held that the wrong done to an individual by (for example) fraud or coercion is not just a matter of the discomfort such an act, if known, would cause him (or an average person). Rather, Kantians have argued that such acts fail to show adequate respect for the individual as an autonomous being, discomfort apart. Therefore Lockeans who would employ a Kantian interpretation of the notion of respect for persons and rights must affirm that what makes endangering acts wrong is not merely the uncomfortable psychological states they may cause in others. This would also fit with the Lockean treatment of actual—as opposed to potential—property crimes, for there it was not essential that the property loss was accompanied by any psychological distress.

(2) Let us, then, consider a second proposal, one that makes no essential reference to psychological states, either on the part of the endangerer or the endangered. Intent and fear alike are displaced from the center of the ethical analysis of endangerment, and a new natural right is recognized: as long as one remains within the bounds of one's own property or common property, one has a right against being exposed by the actions of others—even when they, too, remain on their own or common property—to an increase in the probability one will suffer wrongful harm.[29] This seems a natural extension of the Lockean natural right not to suffer wrongful harm to the case where violation is merely probable. If you like, it recognizes one's safety, or freedom from risk of wrongful harm, as part of one's property.

However, such a natural right would impose heavy restrictions upon the free action of others. Do you have a natural right that I not read *Crime and Punishment* if this would cause a smallish increase in the probability that I might one day rob and bludgeon you? Even inaction on my part may add something to the probability you will suffer wrongful harm, e.g., if I fail to speak sternly to a surly youth, perhaps encouraging him down the road to delinquency. Holders of a Lockean view would no doubt object strongly to such a right, pointing out that it would intrude grossly into an individual's proper sphere of action. Yet on Lockean views an individual's proper sphere of action is not an independent concept. It is defined as the area left open by the exercise of natural rights (one's own and others'), and so cannot be used to determine those rights.[30] Now, I am perfectly sympathetic with the claim that to recognize a right against endangerment would be to restrict individual freedom excessively, but Lockeans should be wary of such arguments: they suggest the possibility of trading off individual territorial rights for some other good, namely, freedom. If natural rights are not set as prior constraints upon the pursuit of any good, even freedom, then all rights in the Lockean canon should be subject to reevaluation. The cautious Lockean will retain the priority of natural rights and look for some nonconsequentialist reason for rejecting or qualifying an extremely restrictive right against endangerment. Several possibilities suggest themselves: one might exclude reciprocal risk, or set a threshold of acceptable risk, or introduce a notion of proximate causation; one might make use of quasi-contractual notions such as tacit or hypothetical consent; or one might pursue a strategy using elements of both these suggestions. In what follows, I will review some of these possibilities.

Reciprocal Risk. Special principles regarding endangerment might not be needed within a Lockean scheme if there were a mutual imposition of risk throughout civil society and states of nature. If this were the case, then all wrongdoing involved in endangerment would in effect cancel out, and it might be possible to avoid the problem of redefining boundaries altogether. This would not be a satisfactory resolution from a theoretical standpoint (are we to limit application of Lockean theory to those societies and those times when the imposition of risk is nearly reciprocal?), but it would as a practical matter eliminate the risk-based problem of finding a middle course between extremes of permissiveness and restrictiveness.

Unfortunately, however, the imposition of risk is manifestly not reciprocal. Not only do some face risks owing to the actions of others upon whom they themselves impose no risks, but imbalance in the magnitude of risks imposed is at least as common as balance. This is especially clear in the three cases with which this volume is concerned: ambient air pollution, pollution in the workplace, and side-stream tobacco smoke.

The Lockean might admit that the imposition of risk is not in fact reciprocal, but then say that it should obey a principle of reciprocity. This would allow individuals substantial freedom of action as long as they do not impose risks upon others greater than those imposed upon them. Such a principle would offer an explanation of what is wrong with smoking in public places or the emission of high levels of pollutants by certain industries: the risk imposed upon others is out of balance with the risk experienced at the hands of others. A normative appeal to reciprocity, however, gives rise to problems absent from its descriptive use. It is barely conceivable that one could keep track of all the risk one experiences and regulate the risk one creates so as to apportion it accordingly, matching both agents and magnitudes. Moreover, the principle has the consequence that it would be impermissible to impose any risks upon future generations, powerless as they are to impose risks upon us. Since virtually any course of action we are likely to pursue will impose risks upon future generations, this principle would hardly enable us to avoid excesses of restrictiveness.

Acceptable Risk and Tacit Consent. Complexities of reciprocal apportionment of risk and quandries about future generations could be avoided if a threshold level of acceptable risk could be established. Let us suppose that we can identify a level of risk that people in general find tolerable. We might then posit a natural right not to be exposed by the actions of others to increases above this threshold in one's risk of wrongful harm. This approach would return some of the neatness of the original Lockean view: so long as one stays below the threshold, one is free to act as one pleases without regard to possible effects on others. Moreover, there is a plausible associated notion of respect for persons: to respect a person involves (among other things) refraining from exposing him to unusual ("unacceptable") levels of risk. Since reciprocity is not presupposed, such a principle could apply across generations.

How might a threshold of acceptable risk be fixed? One could simply observe what levels of risk people do in fact accept in their daily lives without taking special precautions or demanding special compensation, and then infer that this level must not be intolerable. In effect, one is assuming something like tacit consent to this risk level.

It might be objected that people notoriously vary in their willingness to accept risks and that we have no business assuming tacit consent to average levels of risk on the part of those who are atypically risk averse. But let us not quarrel about this, for the whole proposal is deeply confused.

The fact that I daily tolerate a level of risk r in no way shows that I am indifferent about whether an additional risk of magnitude r is imposed upon me, yet the right at issue concerns increments of risk. Moreover, if rational individuals accept a level of risk r, we may be sure that is because they feel they gain something in return, if only convenience. Whether such individuals would find objectionable the imposition upon them of some further risk, even if very much less than r, will depend upon whether they receive something worth the risk in return. It makes no more sense to ask for a level of "acceptable risk" in general than for an "acceptable price" in general. Is five dollars an acceptable price? Well, what is it the price of? Something I want? Is the same thing or a good substitute available to me elsewhere at lesser cost? What are the total resources available to me at the time, and what other spending options exist? It would not be rational for me to accept a one-in-a-million chance of harm if it brought with it no possible benefits, or if other alternatives offered lower risk or greater benefits without greater cost. By looking at the choice behavior of rational individuals, then, we do not discern anything like a threshold of significance with regard to risk; risks are accepted relative to a particular range of options, with an eye to possible benefits.[31]

Suppose, however, that sense could be made of the idea of taking a certain degree of endangerment as a threshold of significance. Would the Lockean's problems be solved? Let us call risk or increments of risk below this threshold level trivial. For simplicity, let us ignore questions about benefits, and imagine that the Lockean arrives at the following principle:

one has a natural right that others not cause a nontrivial increase in the probability that one will suffer wrongful harm as a consequence of their actions.[32] Like other Lockean rights, this one would be quite restrictive regarding polluting activity; just how restrictive depends upon what counts as a trivial degree of risk.

Yet it is not clear how such a right would be deployed within a Lockean scheme, for Lockean rights characteristically apply between individuals. Consider the following sort of case. A polluting act by one individual spreads a toxic substance over a large area. Each of the individuals in that area suffers only a trivial increase in the probability of suffering a wrongful harm as a result, so no individual rights against endangerment are violated. But the probability that *someone* in the area will be wrongfully harmed may be nontrivial—indeed, it may be arbitrarily close to one—suggesting that the act should be impermissible even though it violates no individual's Lockean rights. In another sort of case, a number of people act separately in ways that each cause a trivial increase in the probability I will suffer wrongful harm. But the result is a nontrivial increase in the probability I will suffer wrongful harm from someone (though not from any one person in particular). Again, no individual violates another individual's rights, but nontrivial risk—indeed, arbitrarily great risk—has been imposed. In an extreme case of this kind, two independent acts, each imposing trivial risk on its own, together produce almost certain harm, as when two pollutants, individually not very toxic but in combination lethal, are independently released in my neighborhood.

Other things equal, a rational person would be just as disturbed at the prospect of suffering wrongful harm at the hands of two independent agents as two acting in conspiracy, or at the hands of someone (he knows not whom) as at the hands of a particular individual. When we apply the Lockean framework to questions of social policy as well as individual conduct, it becomes still more obscure why it should matter whether a nontrivial increase in risk is due to the act of one individual or two, or whether an effect will be borne by individuals as a group rather than singly. The image of individuals holding rights against individuals, and of individual trespass as the paradigm of impermissible action, ceases to be illuminating. It would seem appropriate to depart from classical Lockeanism enough to take into account aggregative effects of endangering behavior, but there is no obvious extension of individualist natural rights theory to cover such cases. (We encountered similar difficulties handling aggregative effects in discussing Lockean views about appropriation from the commons.)

The very use of a notion of acceptable risk has already called for a significant departure from classical Lockeanism, for the notion of a threshold of "significant effect" has no answer in traditional Lockean property rights.

The consistent Lockean must explain why it is morally impermissible for me to commit trivial theft, but permissible for me to endanger others to a small degree. Of course, he may say that deliberate or negligent endangerment is always wrong, regardless of degree—the threshold applies only to innocent endangerment. However, in the case of private property, Lockeans have held that it is (objectively) wrong for me to take something belonging to another even if I do so innocently. This disparity in the treatment of innocent endangerment vs. innocent misappropriation could not be justified by a Lockean on the grounds that it is socially efficient to enforce laws against theft rigorously while permitting some latitude when it comes to endangerment. This may indeed be an efficient arrangement, but we are here concerned not with civil codes but with natural rights, which cannot be overridden or abridged—on a Lockean view—for the sake of efficiency. Nor can a Lockean justify the disparity by pointing to our general social tolerance of low levels of risk; we also tolerate low levels of theft.

The notion of an acceptable level of risk thus proves both dubious in itself and difficult to render consistent with a Lockean natural rights theory. There may be some hope for a natural right based upon a threshold of risk, but only if some imposing problems can be solved in a Lockean spirit, and I see no such solutions in the offing.

Causal Proximity and Complexity. I burn some coal to heat my house; sulfur compounds released in combustion enter the atmosphere; in time, these compounds are picked up by water droplets in the clouds, forming dilute sulfuric acid; these acid droplets then rain down on the surface of the earth, slightly blighting your health and home. A Lockean might note that tort law embodies criteria of proximate causation that lessen or remove liability for certain highly mediated outcomes of one's acts. By emphasizing the indirect character of many pollution-caused harms or risks, a Lockean might be able to find elbow room within his scheme. One would be obliged only to refrain from those acts that would proximately cause (or threaten to proximately cause) wrongful harm.

Intuitively, there is great appeal to such a suggestion, but I suspect that much of the appeal comes from what has by now become a familiar confusion: mistaking the question "How much responsibility should we assign to an individual for a given harm?" for the question "Other things equal, is it right or wrong in an objective sense to initiate a chain of events resulting in harm to an innocent?" In the former case, but not the latter, length and complexity of causal chains seem potentially relevant; yet it is the latter that concerns us. Suppose that, to scare a crow away from the soup pot at our campsite, I pitch a rock at him, which ricochets off the pot, strikes a tree branch, rolls down the side of a tent, and drops squarely into the mouth of a sleeping fellow camper, chipping his expensive bridgework. Clearly, if I could have foreseen the whole sequence, I would have been obliged not to toss the stone, i.e., I objectively ought not to have performed the act. Would this judgment be altered in the least if a few more steps had occurred between my act and his harm? One may of course think I deserve less punishment than someone who deliberately took aim at his dozing friend, but even here intent matters more than directness: if I *had* been aiming at my friend, but my arm was unreliable, the stone might have ricocheted off the pot, struck a branch, . . . and I would be as culpable as if I had had a more accurate arm. Length and complexity of causal chain have much to do with foreseeability, and therefore with assessments of intent or negligence, but long and involved causal chains that terminate in wrongful harm are not more objectively permissible than short, straightforward ones, other things equal.[33]

Consent, Hypothetical Consent, and Compensation. Perhaps it is time to stop casting about for an appropriate way of avoiding or weakening a Lockean natural right against endangerment. After all, the Lockean has at his disposal a device for achieving great flexibility in restriction and permission even if the right is absolute: consent. Individuals are entitled to exchange, sell, or give away their rights against endangerment. Through the arrangements made among individuals, the Lockean scheme makes a place for trade-offs of rights against benefits. To carry justificatory force, such consent would have to be free and undeceived (at least, the deception could not have come from the other parties to the agreement). Must the consent also be informed? (*How* informed?) Rather than take up these issues, let us suppose that the Lockean has devised an acceptable account of the criteria of legitimate consent.

Arguably, there are cases in which free and informed consent has been given, yet one ought not to perform the consented-to act. For example, if the agreed-upon terms of a contract prove quite onerous, it may be that I should release the other party (perhaps in return for some compensation) even though I have much to gain by holding him to it. Agreements have a way of becoming onerous even when entered into with what was at the time good information and reasonable care; one can simply be exceptionally unlucky with what was a rational, calculated risk. When this occurs, it may be unconscionable to hold someone to a conscionable contract. In another kind of case, a smoker may find it rare for anyone to refuse his request for permission to light, yet it may still be wrong in some such cases for the smoker to exploit this reticence. It is hardly a new idea that there may be obligations to others not based upon rights. I may, for example, be morally obliged to help you if you are in need and the necessary aid would not be burdensome to me, but not because you have a right to my assistance. Mightn't there also be humanitarian obligations not to cross voluntarily opened borders in some circumstances? Such speculations may not be very libertarian in spirit, but a reasonable notion of respect for persons as ends would seem to involve some humanitarian obligations of this kind. Thus, a Lockean scheme seeking to express such a notion may be unable to treat consent—even when free and informed—as an unproblematic source of justification for the imposition of harm or risk.

Let us ignore such problems, and accept for now the view that individual sovereignty includes the sovereignty of consent. Certainly, it is part of the attractive antipaternalism of Lockeanism that individuals are treated as the ultimate judges and guardians of their own interests.

Even so, it may not be feasible to make widespread use of explicit consent to gain flexibility regarding endangerment. Someone contemplating an activity that involves releasing harmful substances into the air seldom is able to confine the risk thereby imposed to those individuals he has been able to consult fully in advance. And a single individual, even though only marginally affected, would be entitled to veto any such activity, however much it may benefit others, by exercising his right against endangerment. Moreover, some of those put at risk may yet be unborn, and no existing individual can bargain for the claims of future individuals.

If it is in practice often impossible to obtain explicit consent to endangerment, a Lockean may propose as an alternative a scheme of after-the-fact compensation to those exposed to harm or risk.[34] If a polluting activity harms an individual, the compensation required would be such that the victim would have been indifferent before the fact between not suffering the harm at all and suffering the harm but receiving the compensation given. If a local factory blackens my house, and the factory pays to have it repainted and provides me with a small sum to cover inconvenience, I may end up as pleased with this outcome as I would have been had the blackening not occurred. If an individual is simply put at risk, the appropriate level of compensation would be the premium one would have had to pay prior to the exposure to make him indifferent between being exposed to the risk (and receiving the premium) and not. The premium need not be a sure thing; individuals may prefer a state of affairs in which they face increased risk of some harm but at the same time enjoy a higher probability of receiving some benefit. When this standard of compensation for harm or risk is used, one is in a sense obtaining hypothetical consent to a package of harm or risk plus compensation: the package would have been acceptable if offered. Such a scheme has some practical advantages over requiring explicit consent before the fact. First, future generations cannot be consulted, but they can be compensated (if, e.g., this generation leaves behind substantial benefits to offset the risks we bequeath to others). Second, one gains flexibility, for in some circumstances it may be more manageable to provide compensation after the fact than to seek to obtain consent beforehand. Whenever compensation would cost less than the amount to be gained by the endangering act, this flexibility would permit a gain in overall efficiency, for once compensation has been made, no one will be worse off, and at least some will be better off, than otherwise would have been the case.

There are, to be sure, limits upon compensability. If an individual dies or loses something irreplaceable, no after-the-fact benefit could compensate him for the harm done. Since many of the injuries we must consider in discussing pollution—whether as actual harms or as things at risk—are of this sort, the scope for avoiding restrictive prohibitions through a system of compensation is much reduced.

Moreover, it may in many cases be no more practical to determine and distribute required levels of compensation to a diverse—and often future—population than to seek their consent. It might be possible to fix on some sort of average level of compensation for broad classes of polluting activities and broad categories of affected populations. But where in the Lockean scheme is there room for the idea that undercompensation to an individual, or obliging an individual to overcompensate, is permissible if it would be inefficient to determine the actual level of compensation needed? Yet polluters, even small-scale polluters such as smokers, generally cannot keep track of everyone they have harmed or put at risk, of the magnitude of harm or risk in each case, and so on. Is one to imagine a smoker passing out nickels to those who ride with him five floors in an elevator, dimes to those riding ten floors, etc., with double pay for those with weak hearts? Would it really be a lessening of restrictiveness if individuals were required to bear such a burden of monitoring effects, determining compensations, securing compensation from those who resist paying, and so on? More sensible would be (for example) a scheme of taxation on tobacco, with benefits paid out to nonsmokers. Such a scheme would lead to much over- and undercompensation, but probably not more than individual efforts.

A more fundamental difficulty with any scheme for compensation is that it runs afoul of a fundamental motivation of Lockean views: their antipaternalism. Even if I reimburse you after the fact for damages I cause to your person or property, I have failed to respect your right to have the last say over what becomes of both. Nor is this problem removed if I compensate you as well for any damage to your self-esteem. Any attempt to use a notion akin to hypothetical consent within a Lockean framework must confront the fact that even though someone might have consented to an arrangement C were he rational, well informed, and so on, we are not letting *him* decide if we do not actually ask his leave and simply impose *C* upon him. How can it be said that *C* is "imposed" upon the individual if he is able to say what would count as compensation? In any practicable scheme of compensation, the injured individual could not be allowed final say in determining the appropriate level of compensation, for this would permit exorbitant after-the-fact demands. Instead, some interpersonal means must be found to determine appropriate levels, one consequence of which would be that individuals—if they are to be compensated at all—may have to accept levels of compensation they would not agree to.

It would seem to be a clear case of using someone—not necessarily misusing him, but using him—if I were to harm or endanger him in order to pursue my own interests, but then made sure to provide after-the-fact compensation at a level he would have to take or leave. Why is this something less than treating him as a Kantian end-in-himself? Several elements seem to be involved. First, there is a preemption of his actual will and of his sovereignty, his entitlement to decide certain matters himself. Second, such preemption reflects an attitude according to which what matters is that people receive certain outcomes, even if they did not participate in bringing the outcomes about through an exercise of autonomous choice. Third, in the simplest sorts of cases, compensation really is nothing but a price attached to the pursuit of one's own ends, a toll one must pay in order to get on with it, a fee that frees one from the obligation of consulting others. Nothing in the compensation mechanism itself prevents one from taking this instrumental view of others, and much encourages it. Finally, ability to compensate will vary with ability to pay, so that those with greater resources will gain greater release from restriction than those with less. If in a Kantian "kingdom of ends" we are all equal and the ends of others are equally our ends, then we seem a long way from such a kingdom in a world in which the better-off are able to preempt the wills of the less well-off through the mechanism of compensation, but not conversely. A rich man may be able to ride his hounds over a peasant's land and then make this up to the peasant monetarily, but in so doing does he show respect for the peasant's property rights, or for the peasant as an equal in the kingdom of ends?

We must make a distinction. Once an unconsented-to harm of endangerment has occurred, is it a respectful thing to compensate? Probably so. But is it a respectful thing to harm or endanger without consent, perhaps even deliberately, so long as one later compensates? Probably not. From a social standpoint, it is, we have seen, efficient to permit boundary violations whenever the violater can compensate his victims and still come out ahead. But one of the central themes of the Lockean tradition has been that individual rights take precedence over efficiency.

Now, it seems to me quite important that it be at times morally permissible to pursue activities that give rise to unconsented-to harms or endangerment when these activities in the end yield a substantial balance of benefits over burdens. Without this possibility, society could not achieve much by way of development. Yet moral theories of a Lockean structure and Kantian inspiration tend to exclude such quasi-utilitarian balancing as disrespectful of persons. Now, the balancing involved in a scheme of violation-and-compensation is not strictly utilitarian, for it is person-specific, requiring that a surplus of benefits over burdens accrue to the particular individuals bearing the burdens of harm or endangerment, and not merely to society as a whole. Yet Kant tells us not to use others solely as means, even as means to their own ends, and libertarians tell us that we cannot force things upon people even when they themselves are the beneficiaries. A scheme permitting trade-offs of benefits against rights, even when person-specific, is thus in important ways conceptually closer to utilitarianism than to either Kant or classical natural rights theory.

It is instructive that Lockeans typically have not advocated a scheme of violation-and-compensation as a way of loosening the restrictiveness of private property in general. For example, one might avoid the need for "excessive governmental regulation and enforcement" of prohibitions against theft or assault by permitting the thief or assailant to compensate afterwards, or, if compensation is not forthcoming, allowing the victims to have recourse to the tort system.[35] Instead, most Lockeans have advocated direct state enforcement via criminal law of rights against theft and assault, and for good reasons: a system of violation-and-compensation would place the burden on the victim; it would allow preventable crimes to occur; it would fail whenever compensation is not possible owing to the nature of the harm or the resources of the harmer; it would be unreliable (many would fail ever to receive compensation because they are unable to pursue their cases, or because small losses would not be worth pursuing); it would not be an adequate deterrent to crime and so would increase insecurity; it would give rise to "free-rider" problems; and so on.[36]

These reasons apply with equal force to the environmental case. In fact, it may in general be harder for individuals to detect and assess pollution-caused injury or risk than is the case in ordinary crimes against person or property. The information-gathering burden on victims would be enormous, substantially diminishing the probability that polluters would voluntarily compensate (since they could often hope to escape detection) or be brought successfully to trial for failure to compensate. Once one makes a serious assessment of what it would involve for individuals to keep up with current knowledge of the effects of pollutants and to trace the origin of the pollutants to which they are exposed, it becomes obvious that a system of regular public regulation, monitoring, and policing would secure substantial gains in efficiency over individual enforcement through threat of suit. Moreover, such governmental activity need not violate any Lockean rights, for no one is entitled to harm or endanger, with or without permission.[37]

Both schemes—direct enforcement by the state and individual enforcement through violation-and-compensation—would require that it be permissible under some circumstances to enter private property to monitor potential hazards. How else could one determine whether another is putting one at risk by acts carried out entirely on his own property? The entitlement to inspect would require that property rights be retailored, just as they are tailored in civil law by the notion of a reasonable search. Government enforcement would have the advantage of limiting the total amount of inspecting activity needed by reducing duplication: government inspectors, but not private individuals, could be required to make public record of their findings. Moreover, public inspectors could be required to carry identification and to follow certain procedures.

Note that if the Lockean picture were modified to demote private property to the status of common property, compensation without consent might be fit in more consistently. Under this modification one could appropriate another's property without consent so long as one left enough and as good either in his remaining private holdings or in readily accessible common property. If a given appropriation would not leave enough and as good in this sense, it might still be permissible if one were to substitute something of value equal to whatever is taken. (Would one also have to compensate for distress?) Compensation, then, would emerge as a special case of meeting the conditions necessary for appropriation.

A Lockean may use compensation to gain lessened restrictiveness either by making special provision for a scheme of compensation without consent, or by treating private property on a par with common property. Both are major changes and would result in moral theories with different, and almost certainly less, intuitive appeal than the orthodox view. They appear to fall between two stools, being defective both with regard to social efficiency—and so not attractive to utilitarians—and with regard to respect for persons—and so not attractive to Kantians.

Lockeans, then, do seem to face a genuine dilemma. The orthodox view turns out to be vastly restrictive of individual freedom when it comes to pollution-caused harm, but insufficiently restrictive when it comes to pollution-caused risk. Revisions of the orthodox view may permit a more sensible balance, but involve significant departures from Lockeanism and bring with them a host of new problems.

Summary and Conclusion

The injuries and endangerment we may experience from pollution have excited among Lockeans an ingenious interest in incorporating greater flexibility into their view, an interest much less commonly directed at the restrictiveness of the system of private property in general. After all, the system of private property calls for centralized, direct governmental regulation, yet it is viewed by Lockeans as the very bulwark of freedom. What structure of governmental permissions and prohibitions should be in place regarding pollution is a subject beyond the scope of this chapter. My point is the modest one that it seems hardly defensible to treat airborne harms differently from handborne harms. I cannot here canvass all the ways a Lockean might attempt to come to terms with risk and injury from pollution, but we may draw some tentative conclusions.

First, if we treat injuries due to polluting activities comparably with injuries that happen to be caused by other means, we find that the Lockean view in its classical form is highly restrictive about pollution, very much the opposite of the laissez-faire doctrines associated with it. It should be viewed as no more probable that problems posed by pollution could be handled by a self-regulating market than problems of property crime in general. Much of the appeal of Lockean views is that they seem to afford a way of securing considerable freedom of action for individuals, but this is so only if we disregard the injuries individuals may cause each other through the medium of the environment. We cannot argue that Lockeanism is internally inconsistent if it turns out that this doctrine would, if put into practice, be very restrictive of freedom of action, for Lockeanism does not require that freedom should be maximized. But Lockeanism will lose much of its attractiveness unless there is good reason to think that a society founded upon Lockean principles would permit very substantial freedom of action. Owing to environmental effects, Lockeanism would, if put into practice, impose much more severe restraints upon individual action than, for example, the most elaborate existing environmental laws and regulations.

Second, if we look at the specific issue of risk, as opposed to actual injury, we find that classical Lockeanism may fail to be restrictive enough, especially if it is to be thought of as giving expression to the Kantian notion of respect for persons.

Third, there are some systematic—but unmotivated—asymmetries in the classical Lockean view, most notably with regard to the treatment of ownership rights in common property vs. private property.

Fourth, the search for a Lockean scheme that strikes a more appropriate balance between restrictiveness and permissiveness suggests a number of modifications of the classical theory: one may need to question the absoluteness of certain rights; to introduce considerations of balancing benefits and burdens; to contemplate collective or aggregative entitlements or obligations as well as individual rights; to challenge the idea that so long as one operates within one's own boundaries and intends no harm, one need acknowledge no other-oriented constraints or obligations; to recognize limitations on the justifying role of consent—tacit, explicit, or hypothetical; to rethink the Lockean notion of privacy; and to give a fuller account of the notion of showing respect for persons.

All this is a rather roundabout way of arguing what perhaps cannot be argued more directly: if we take seriously the fact that we find ourselves situated in, and connected through, an environment, we are soon impressed with the inaptness of a conception of morality that pictures individuals as set apart by propertylike boundaries, having their effect upon one another largely through intentional action, limiting their intercourse by choice, and free to act as they please within their boundaries, although absolutely constrained by them. The result of this conception is gross restrictiveness here, gross latitude there, and, in general, an inadequate vocabulary for debating, or even expressing, a number of pressing moral issues concerning the environment.

It does not follow that one ought to give up the notion of individual natural rights as the basis of morality and adopt utilitarianism. But the arguments made here at least suggest that a plausible morality will involve at base more than a scheme of presocial, territorial individual rights and will make room for a number of notions—balancing, aggregation, and the like—more commonly associated with utilitarian than natural-rights theories.[38]

Notes

I am grateful to members of the Working Group on Risk, Consent, and Air for their comments on an earlier version of this chapter. I should mention especially Judith Jarvis Thomson, Samuel Scheffler, Mary Gibson, and Douglas MacLean. I would also like to thank Rebecca Scott for much helpful discussion.

1. I do not attempt here a full characterization of the doctrines of John Locke. Rather, I seek to draw out certain main features of an influential view that takes a number of central Lockean doctrines as its foundation. For example, although contemporary libertarians often draw heavily from Locke, it would be misleading to call Locke himself a libertarian.

2. See John Locke, *The Second Treatise of Government* (Indianapolis: Bobbs-Merrill, 1952), Chap. 5. All references to Locke are to this work.

3. At this point in the argument, Locke has assumed the existence of an imperishable medium of exchange; otherwise, any wealth amassed beyond what one could use would perish and be wasted. Locke, sec. 51.

4. Perhaps the most explicit use of this boundary-based image of moral space is in Robert Nozick, *Anarchy, State, and Utopia* (New York: Basic Books, 1974), Chap. 3. All references to Nozick are to this work.

A right in such a scheme is a moral liberty or entitlement to do or refrain as one pleases, and it entails the existence of an obligation on the part of others not to interfere. In effect, then, it establishes a border. One possible exercise of such a right is to grant to another part or all of one's entitlement, i.e., to issue a pass (the border stays put, but the other is free to cross—perhaps subject to certain conditions), to transfer a deed (the border stays put, but the rightful occupant changes), or to establish joint ownership (no property line separates joint owners, but there may be some agreed-upon limits governing use; in effect, if we think of the uses of properties as quasi-spatial dimensions of it, such limitations are internal boundaries).

5. Or, would it involve crossing an internal boundary of jointly owned property? (See Note 4.)

6. A complication arises where another has a "settled design" upon one's life or possessions; then self-defense permits interfering with him even before he crosses one's boundaries. He has, according to Locke, declared an unjust war upon one and thereby forfeited his rights against interference. See Locke, sec. 16.

A futher complication is that in a state of nature Locke gives to all the right to enforce the laws of nature. One may, presumably, cross a boundary for this purpose without asking permission, e.g., to retrieve stolen property and secure as well any of the thief's property needed for "reparation and restraint" (sec. 8). This boundary crossing, too, comes under the head of states of war (Chap. 2).

7. For example, this view is taken by Nozick, pp. 30ff.

8. Hereinafter, when I speak of crossing a boundary wrongfully, I will mean that the crossing was not freely consented to, is not a legitimate act of self-defense, and is not a legitimate effort to enforce the laws of nature.

9. An objective obligation is an obligation one would recognize if one knew all the principles of morality and all the relevant facts about one's situation and drew the appropriate moral conclusion. (It does not follow that one necessarily would be moved by this conclusion.) Subjective obligations are relative to what one believes, perhaps wrongly, to be the case. For discussion of the distinction, see Richard B. Brandt, *Ethical Theory* (Englewood Cliffs, N.J.: Prentice-Hall, 1959), pp. 362ff.

10. Hume wrote that "a poor peasant or artisan . . . [who] knows no foreign language or manners and lives from day to day by the small wages which he acquires" cannot seriously be said to give consent to his government simply by remaining in his place. "We may as well assert that a man, by remaining in a vessel, freely consents to the dominion of the master, though he was carried on board while asleep and must leap into the ocean and perish the moment he leaves her." David Hume, "Of the Original Contract," in *David Hume's Political Essays*, edited by C. W. Hendel (Indianapolis: Bobbs-Merrill, 1953), p. 51.

11. Locke himself held that we could not trade away our right to life or sell ourselves into slavery, this being contrary both to reason and to the will of our ultimate owner, God (secs. 6 and 25). Nozick, on the other hand, imposes no such restriction (p. 331).

12. I assume that your neighbor is not operating the reactor in a Kamikazelike effort to put your life at risk, i.e., has no "settled design" to harm you, and so his behavior does not fall under Locke's special provision for self-defense. The question of intent will be discussed further below.

13. I am availing myself here of the notion of a probabilistic cause. Roughly, a factor C is a (partial) probabilistic cause of event E at time t if, were C not present at t, but conditions up to t were otherwise exactly the same, the objective probability of E at t would have been lower than in fact it was. Thus characterized, this notion is neutral on the question of whether there is an underlying determinism.

14. Interestingly, Locke thought that in a state of nature no one would bother to appropriate in a wasteful fashion, since it would simply be a net loss to expend one's labor and then not put the product to good use (sec. 51). This would be so only if it were true that the simplest way of obtaining something for one's own use did not commonly involve despoiling other parts of common property not put to good use. Historically, the opposite has often been true and special care and effort would have been necessary to avoid wasteful despoilation.

15. Regarding this unclarity, see Nozick, pp. 174–75, where he rather surprisingly leaves the unclarity unresolved.

One common form of pollution, dumping toxic wastes into the soil, can be quite a bit like creating a cloud of pollutants in the atmosphere, although this sort of pollution is usually accompanied by some leaching into ground water, streams, etc.

16. Talbot Page, "A Generic View of Toxic Chemicals and Similar Risks," *Ecology Law Quarterly* 7 (1978): 207–44, 222n. Page cites K. S. Crump, D. G. Hoel, C. H. Langley, and R. Peto, "Fundamental Carcinogenic Processes and Their Implications for Low Dose Risk

Assessment," *Cancer Research* 36 (1976): 2973–79, and Jerome Cornfield, "Carcinogenic Risk Assessment," *Science* 198 (1977): 693–99, p. 696.

17. It should be noted that the requirement of consent before taking private property applies among joint owners. Thus, if two of us were to inherit an estate jointly, I could not simply take whatever I wished from the estate so long as you were left "enough and as good"; some sort of agreement would be needed before the estate could be rightfully partitioned. Express consent may not be needed when tradition or custom establishes rules governing joint property or where the joint owners are in some other, special legal relationship (such as marriage). Locke notices the need for consent when he considers "the joint property of this country or this parish," which he distinguishes from common property in a state of nature by noting that it is "common in respect of some men, [but] it is not so to all mankind." He treats this as, in effect, jointly owned private property, saying that "no one can enclose or appropriate any part without the consent of all his fellow commoners." (Locke adds, as a quite separate consideration, that in practice if one were to appropriate part of such property one would not ordinarily leave "enough and as good" behind in common.) An asymmetry therefore exists between jointly held common property and jointly held private property, even though it would seem as if the difference in the end could be no more than a matter of the size of the joint-holding group: a parish, a country, or mankind. See Locke, sec. 35.

18. There are other problems with Locke's argument. For example, it at best justifies private consumption, but not private ownership of land or of other productive resources that might be put to common use in creating the requisites of private consumption.

19. Perhaps some of the special status often attributed to private property derives from considerations of privacy and from the especially intimate relationship we may have with our own possessions. This is a mire of issues — not the least of which concern the private and intimate relationship we may have with common property — which I will simply refrain from wading into. This is consistent with our purposes, for what is at issue is not a bare right to hold some property privately, but a right of exclusion that can be extended over property in no way intimate to us or necessary for us.

20. There is a third option: to drop the notion of *natural* property rights altogether. This leaves the possibility of arguing for a system of civil property rights (see below) and is compatible with recognizing other natural rights in terms of which such a system might be justified.

21. At least at one point, Locke himself suggests a very loose reading. He considers the "rule of property, viz., that every man should have as much as he could make use of" and says it "would hold still in the world without straitening anybody, since there is land enough in the world to suffice double the inhabitants," had not money been introduced (sec. 36). The criterion that others not be "straitened" is clearly weaker than that they be left enough and as good. Elsewhere, Locke says that "enough and as good" means "there was never the less left for others" and that "he that leaves as much as another can make use of does as good as take nothing at all" (sec. 33). These latter remarks suggest a stricter and perhaps more appropriate standard: that no one's prospects be reduced by another's appropriation.

22. Again, we assume that it is not part of his motive to put you at risk by operating an unsafe reactor.

23. Depending upon details of particular Lockean theories, it may be that extreme gains or losses in social utility or in consequent rights observance would offset violations of individual rights. Nozick, for one, says the question of absoluteness "is one I hope largely to avoid," p. 30n.

24. The expression "side constraint" and the image associated with it are found in Nozick, pp. 28ff.

25. Locke, *The Second Treatise of Government*, sec. 16.

26. Nozick, in Chapter 4 of *Anarchy, State, and Utopia*, does use the concept of fear in discussing risk, although he does not give a general theory of how the concept of fear would be integrated into his treatment of natural rights. One may also ask whether it is a harm to cause fear even when no actual risk is involved (and when causing such fear involves no deliberate acts of deception, etc.).

27. Someone might say: "But some people are just more sensitive than others, and it would not be fair that they should therefore have more extensive claims upon the rest of us not to be harmed psychologically. They do not *deserve* such sensitivity — they may merely be born with it or have inculcated it in themselves." This would be a peculiar objection in the mouth of a Lockean, for some are born with greater property than others, and some acquire greater property through their own efforts, thereby gaining — justifiably, in the Lockean's eyes — more extensive claims against others than the rest of us enjoy.

In comparison with the acquisition of property, is it simply too easy, and of too little benefit to others, to develop a thin skin? Of course, it sometimes is quite easy to acquire property (e.g., by inheritance) and quite unhelpful to others, but that aside, it might be thought that permitting individuals to enlarge their sphere of moral claims through the acquisition of property provides an incentive to industriousness, which often benefits others as well as the agent. By contrast, it is hard to see what beneficial incentives would arise from allowing individuals to enlarge their sphere of moral claims through the cultivation of exquisite sensitivity or the manufacture of psychological harms. This is plausible enough, but unfortunately for the Lockean it is a straightforwardly utilitarian argument.

28. Could it be said that someone is *helping* you by tipping you off, that is, giving you grounds for rational fear of wrongdoing on his part if his plan is in fact to harm you? In those cases where you are able to put this information to good use, there is some benefit along with the (perhaps protracted) anxiety. However, it is a general feature of harm-causing acts that they may in certain circumstances also confer benefits, so we cannot use this point to settle the question whether causing rational fear of wrongdoing on one's own part is a harm.

29. Does the reference to probability reintroduce psychological states, in the form of degrees of belief? Not necessarily. First, the probability in question could be an objective probability, such as a propensity or a relative frequency, and thus be fully independent of what the agents in question believe. Second, for those who do not admit of irreducible objective probability, it could be interpreted as an idealized subjective probability — a rational degree of belief conditioned upon all relevant evidence. This is manifestly not a psychological state of any sort as such — such probabilities exist at a time whether or not anyone thinks of them.

An important question hinges upon the interpretation of probability given here. Is it objectively wrong, for example, to introduce a drug that in fact is harmless but for which there is not adequate evidence to warrant the conclusion that it is harmless? That is, is it objectively wrong to act in a way that could be said to increase the subjective probability another will suffer harm even if no increase in objective probability actually occurs? My inclination is to say that what is wrong in such cases is that one is acting contrary to one's subjective duty, although, fortuitously, this turns out not to be contrary to one's objective duty.

30. On this point, see the otherwise quite baffling definition of "voluntary action" given by Nozick:

> Whether a person's actions are voluntary depends upon what it is that limits his alternatives. . . . Other people's actions place limits on one's available opportunities. Whether this makes one's resulting action non-voluntary depends upon whether these others had the right to act as they did. [p. 262.]

On this account, I leave your home voluntarily if you tell me (which is within your right) that if I do not do so you will call the police.

While this clearly is inadequate as an account of voluntariness, one can see the Lockean forces that drove Nozick to this position: whether an act is within one's proper sphere of free action depends upon whether the constraints others impose upon it are within their right to impose. However, not all acts within one's proper sphere of free action need be voluntary; whether they are depends as well upon the psychology of the agent. Moreover, acts outside one's proper sphere of free action may be quite voluntary, albeit wrong.

31. Further, if it is rational for individuals to assess costs vs. benefits in evaluating the acceptability of risks, this seems equally rational at the social level. Yet the Lockean view precludes social aggregation and balancing.

It does not follow that so-called "cost-benefit analysis" is a uniquely rational way of making social policy. Indeed, it does not follow that cost-benefit analysis is even minimally rational, for it suffers the following defects (among others): it fails to take into account the declining marginal utility of money (and of other ways in which distribution may affect utility at a given cost/benefit ratio); it ignores the disparities between utilities and prices (including the fact that prices, but not utilities, reflect ability to pay, and that future utilities cannot be discounted the way future prices can); and it often substitutes the demonstration of a net surplus of benefit over cost for a demonstration that a given course of action is optimal. For further discussion, see my "Costs and Benefits of Cost-Benefit Analysis: A Response to Bantz and MacLean," in *PSA 1982*, Vol. 2 (East Lansing, Mich.: Philosophy of Science Association, 1983).

32. A similar principle is suggested by Judith Jarvis Thomson in Chapter 6 of this book, "Imposing Risks."

33. I am grateful to Judith Jarvis Thomson and David Lewis for discussion of the possible importance of a causal proximity condition.

34. One such scheme is Nozick's. I follow his account in most details. See Nozick, *Anarchy, State, and Utopia*, Chap. 4.

35. This in fact parallels proposals one hears from libertarian quarters for dealing with pollution. The parallel is rather close in the case of Nozick. Just how close is hard to say, owing to loose ends in Nozick's account and to my imperfect understanding of the whole of his view.

36. Would anyone be interested in perpetrating aboveboard theft-and-compensation? One might dearly love to gain use or possession of a particular piece of property one does not happen to own and be prepared to compensate the owner adequately after the fact, but for one reason or another be unwilling or unable to obtain express consent to such an arrangement. Indeed, if one could simply make more efficient use of certain property than its present holder, one might be able to take it without permission, use it, pay full compensation (i.e., give the owner the equivalent of what he would have had if he had kept possession of the property), and still come out ahead. (See the discussion, below, of demoting private property.) Aboveboard *assault*-and-compensation may have substantial appeal to various individuals for reasons we need not go into here.

37. Some Lockeans would insist that individuals remain free to sell or waive their rights against harm or endangerment, just as they should be free to sell or waive rights against assault or theft.

38. My sense of fairness forces me to note that a difficulty mentioned above poses a problem for those utilitarian theories that assess the objective rightness or wrongness of acts in terms of the value of the *actual* consequences they produce. (Those utilitarian theories based upon the *expected value* of acts will not have this problem.) The existence of an unactualized possibility of harm will not show up among manifest consequences, but this means such theories will not reflect pure risk as such. Two acts with the same manifest consequences, but differing in that one imposes a substantial (but unactualized) risk while the other does not, would be judged morally equivalent, other things equal. This is somewhat counterintuitive. Consider a closely related problem. Suppose that an act of mine substantially augments the market value of your house, that is, increases the price that would be paid for it were it to be sold. But suppose further that this increase in value is only temporary, that you do not sell during this period, and, indeed, that you never learn of this change in value. Have you benefited from my action? (Does one benefit by receiving a lottery ticket, hopefulness aside, only if it wins?) If this is a benefit, then we should count possibilities, not just actualities, among the valuable consequences of actions—even possibilities that are never actualized. That would allow a consequentialist to capture the intuitive judgment that, other things equal, it is objectively worse to impose more rather than less risk, even if the risk remains pure. Of course, we would need to explain why mere possibilities are benefits (or harms) and to ask whether, for example, any amount of merely possible benefit (or harm) could outweigh even the smallest actual benefit (or harm). I leave these puzzles for another occasion.

6

Imposing Risks

Judith Jarvis Thomson

1. I think it pays to distinguish three kinds of case. In the first kind, an agent causes an unwanted outcome by his act, or by each of a series of acts. In the second kind, the agent causes an unwanted outcome and imposes a risk of a further unwanted outcome. I shall call cases of the second kind cases of "impure risk imposition." In the third kind of case, each time the agent acts he imposes a risk of an unwanted outcome, and it may be that he never at any time actually causes an unwanted outcome. I shall call cases of the third kind cases of "pure risk imposition."

2. B owns some tomatoes; A steals one; A thereby causes an outcome unwanted by B, viz., loss-of-tomato. B also owns some carrots; A steals one; A thereby causes another outcome unwanted by B, viz., loss-of-carrot. Each time A steals a vegetable from B, he causes an outcome unwanted by B. This is a case of the first kind.

I shall leave the limits of "unwanted outcome" dark. I mean to include in the range of unwanted outcomes various kinds of harms to persons or their property, a harm being a more or less long-lasting injury. (I include death itself among the harms.) I also include pain and some cases of discomfort. (If you pinch a man's nose, you may cause him pain, but no harm, there being no injury caused him.) I also include some cases of fear. I also include some cases of nuisance. Thus if I open a glue factory in my house, I thereby cause my neighbor an unwanted outcome in the sense of this chapter. By contrast, if I decorate my front lawn with concrete gnomes and plastic geese, I certainly do something my neighbor wishes I had not done, but I do not cause an unwanted outcome in the sense of this chapter.[1] Perhaps the difference lies in the pervasiveness of the nuisance in the first case. My neighbor can escape the smell of my factory only with difficulty, and at considerable cost to himself, whereas my neighbor can escape the sight of my gnomes and geese more easily (e.g., averting his eyes as he goes in and out, planting shrubs between our lawns, etc.). Or perhaps something more or else is involved. We do think of a dislike of certain kinds of sights as a matter of taste; we do not think of a dislike of certain kinds of smell as a mere matter of taste. It is an interesting question why, but I simply pass it by. I shall make no attempt to give a precise characterization of what counts for present purposes as an unwanted outcome. It may help, however, to say that what I have in mind is very roughly characterizable as follows: other things being equal, to cause a person an outcome of the kind I mean is to infringe a right of his.

A particularly interesting subclass of the cases of the first kind is the subclass of cases which involve "threshold effects." Suppose I put a cupful of poison in your pond—not enough to kill the fish, but enough to cause the water to turn muddy-looking. I have caused you a harm. Now I again put a cupful of poison in your pond. Still the fish live, but now the water smells bad, another harm. I put the same amount of poison in the third time, and lo, the fish all die, a very serious harm to you. Each act in the series causes a harm; the last act causes a far more serious harm than the preceding members, but would not have caused that far more serious harm if it had not been preceded by the earlier members of the series. Cases of the kind I have in mind here, then, are cases in which earlier members of the series each cause a harm, and the cumulative effect of those harms is such that a later member of the series causes a far more grave harm, in the circumstances, than it would otherwise have caused.

I think this subclass of cases of particular interest since I suspect that anyway some of the cases of pollution which concern people are cases of acts, or series of acts, which fall into this subclass.

In any event, it is worth taking note of the fact that the cases we have been looking at so far raise no worry about imposition of risk. Each act in every series of acts which falls into the first kind of case is an act which, itself, causes a harm, and it is the harm it actually causes which gives the victim his ground for complaint against the agent. Insofar as risk is involved, the risk is that the agent will continue to act, i.e., go on causing harm, and eventually perhaps cross a threshold, causing a major harm. The risk, as it were, lies between agent and act, not between act and harm. No special moral considerations raised by imposing risks on people need be attended to in order to settle whether it is morally permissible for the agent to proceed.

3. Cases of the second kind are mixed. Suppose A shoots B in the stomach. Then A causes B a very serious harm; but he also imposes a risk of a still more serious harm, viz., death. Before the shot, B's chances of dying within the week were (let us suppose) very small, since (let us suppose) B

was young, in good health, etc. After the shot, B's chances of dying within the week are high because the stomach wound may cause shock, may become infected, etc. B has a twofold complaint against A; he has ground for complaint against A in that A caused him a serious harm, and he has yet another ground for complaint against A in that A has imposed a high risk of death on him.

Series of acts may fall into this class too. Each time I smoke a cigarette in the office, I cause Jones discomfort: runny eyes and nose, throat irritation, etc. (He also hates the smell.) But if I also, each time, impose a risk on him of lung disorder, and even death, then he has two grounds for complaint against me each time I smoke in the office: the unwanted outcome my smoking does in fact cause him, and the risk of serious harm I thereby impose on him.

Perhaps there are threshold effects here too. It might be that after I have been smoking in the office over a period of time, my next act of cigarette smoking imposes a considerably greater risk of harm on Jones, in the circumstances, than it would have done had it occurred earlier in the series.

Where the unwanted outcome actually caused is relatively trivial by comparison with the unwanted outcome risked, the outcome actually caused may drop out as appearing to be morally insignificant. The case may then seem to fall into the third kind. That is because the ground for complaint lying in the risk imposed is so much more grave than the ground for complaint lying in the unwanted outcome actually caused.

4. Cases of the third kind are pure. Suppose A played Russian roulette on B. B has ground for complaint against A even if B was caused no harm (no bullet was under the firing pin when A fired), and even if B was unaware of what happened, so that he was caused no fear or discomfort. The ground for complaint lies in the fact that A imposed a risk of death on B.

If A pours poison into B's fish pond, A causes B a harm. Suppose, instead, that A's smokestacks emit sulfur dioxide, which then combines with water so that acid rain may or may not fall into B's fish pond. In the latter case, A may or may not cause B the harm which he causes in the former case; in the latter case, then, A's acts may merely have imposed a risk of harm.

I shall concentrate on cases of this kind, since they are the ones in which the role of the imposition of risk in moral assessment of action is clearest.

The central problem which pure risk imposition raises for moral theory is this: which instances of pure risk imposition are wrongful, and which are not. The central problem which pure risk imposition raises for political theory is this: which kinds of pure risk imposition should be made illegal, and which should not. I am able to do no more than make some remarks about one of the difficulties which I think gets in the way of an attempt to solve the central problem which pure risk imposition raises for moral theory.

5. Suppose that if A does such and such now, he will thereby impose a risk of harm on B. (I shall from here on use the term "harm" to cover all of the unwanted outcomes.) Is it permissible for him to do so? Or would it, instead, be true to say to him "You ought not"?

I should perhaps say straightway that I shall throughout use "It is permissible for so and so to do such and such" as an equivalent of "It is not the case that so and so ought not do such and such."

Some such cases seem to be easy to deal with. If the circumstances are such that A would be justified in causing B that harm, then surely A would be justified in imposing a risk of that harm on B, i.e., it would not be true to say to him "You ought not." Suppose, for example, that B is coming at A with a knife, obviously meaning to kill him. And suppose that A has no other way of preserving his life than by killing B. Then, other things being equal, it is permissible for A to cause B's death. It seems to follow that it would be permissible for A, in those circumstances, to impose a risk of death on B.

I said that some of these cases seem to be easy to deal with, but I do not for a moment mean to imply that it is easy to say under what conditions it is permissible for one person to cause a harm to another. Cases of self-defense are relatively easy (but only relatively easy); but in other cases the matter is much less clear. For example, aren't there cases in which it is permissible for A to cause a harm H to B in that doing so is the only way in which A can save C and D from suffering a considerably graver harm G? But which cases are they? The question what in general are the conditions in which causing such and such a harm is permissible is one of the central and most difficult questions in moral theory. My point here is only this: other things being equal, it is permissible for A to do something which will impose a risk of harm H on B if it is permissible for A to cause B harm H. That, I think, seems plausible enough.

If we could also say "only if" with equal plausibility, then it would be plausible to think that risk imposition generates no independent problem for moral theory. Any question whether it is permissible to impose a risk of harm would be reducible to the question whether it is permissible to cause that harm. We would be free to ignore risk imposition and concentrate on the (exceedingly difficult) question under what conditions it is permissible to cause this or that harm.

But there unfortunately is what looks like good reason to think we cannot also say "only if." For example, my neighbor is not now coming at me with a knife. (It is early morning, and he is still asleep.) Nor is there anybody whose life or limb I can save by causing my neighbor a harm. It certainly seems plausible to think that the circumstances which now obtain just are not circumstances in which it would be permissible for me to cause my neighbor any harm at all; a fortiori, it seems plausible to think that if anyone said to me now

(1) You ought not cause your neighbor's death,

he would be speaking truly. So far so good. In fact I want some coffee now, and must turn my stove on if I am to have some. If I turn my stove on, I impose a risk of death on my neighbor—it is a gas stove, and my turning it on may cause gas to leak into his apartment, or it may cause an explosion, etc. Feeling a surge of moral anxiety, I ask your advice. You say: Absurd. That's a fine stove, in mint condition, and the risk is utterly trivial. So it's not the case that you ought not turn your stove on; i.e.,

(2) It is permissible for you to turn your stove on.

It is very plausible to think that (1) and (2) are true; but if they are, we cannot say "only if." For my turning my stove on will impose a risk of death on my neighbor, and if (2) is true it is permissible for me to turn my stove on; but if (1) is true it is not permissible for me to cause my neighbor's death.

But perhaps this example sailed by too quickly. *Is* (2) true?

Suppose that, feeling reassured, I turn my stove on. Lo—astonishingly, amazingly—my doing so causes an explosion in my neighbor's apartment and thereby causes his death. Question: Does this show that you spoke falsely when you said (2)? That seems to me to be a very hard question to answer.

We know what G. E. Moore would say. He would say that—astonishingly, amazingly—it has turned out that you spoke falsely when you said (2). He would say that I am not to blame for my neighbor's death, or for anything else for that matter, for I had every reason to think the risk trivial, and I did not turn the stove on with a view to causing my neighbor's death, or even in the belief that I would cause his death; I turned the stove on only to make coffee, as I do every morning, in perfect safety. And Moore would also say that you are also not to blame for anything, and that you were justified in saying what you said. For, as he would remind us:

> we may be justified in saying many things, which we do not know to be true, and which in fact are not so, provided there is a strong probability that they are.[2]

It could even be added—though Moore does not add it—that, not merely am I not to blame for anything, but more, the act which consisted in my turning the stove on was not a morally bad act, for I had every reason to think the risk trivial, and I did not turn the stove on with a view to causing my neighbor's death, or even in the belief etc. etc., as above.

One who takes Moore's line does not thereby commit himself to the view that risk imposition generates no independent problem for moral theory; but it is not obvious what independent problem it does generate if Moore's line is correct.

I think we all feel some inclination to take Moore's line. Surely what a person ought or ought not do, what it is permissible or impermissible for him to do, does not turn on what he thinks is or will be the case, or even on what he with the best will in the world thinks is or will be the case, but instead on what *is* the case.

Some people would say that these things are true only of one of the two (or more?) senses of the word "ought": the "objective" sense of "ought." And they would contrast it with a (putative) "subjective" sense of "ought." Presumably the latter (if there is such a thing) is parasitic on the former, i.e., presumably "He (subjective) ought" means "If all his beliefs of fact were true, then it would be the case that he (objective) ought," or perhaps, more strongly, "If *all* his beliefs were true, then it would be the case that he (objective) ought." But I greatly doubt that there is such a subjective sense of "ought." On those rare occasions on which someone conceives the idea of asking for my advice on a moral matter, I do not take my fieldwork to be limited to a study of what he believes is the case; I take it to be incumbent on me to find out what *is* the case. And if both of us have the facts wrong, and I therefore advise him to do what turns out a disaster, I do not insist that in one sense my moral advice was all the same true, though in another sense it was false.

But does rejecting the idea that "ought" has a subjective sense require us to agree with Moore in saying that you spoke falsely when you said (2)?

Why might one think that you spoke falsely when you said (2)? Well, here is one possible route to the idea that you did. At the time at which you spoke there are two possibilities:

(a) If she turns her stove on, she will thereby cause her neighbor's death

and

(b) If she turns her stove on, she will not thereby cause her neighbor's death.

You thought (a) highly improbable, and that is why you said (2). But suppose we make an assumption. What I have in mind is the assumption that if a proposition says that something will happen, and the something does happen, then it always was the case that the something would happen, i.e., it always was the case that the proposition was true. In particular, then, we would be assuming that, since I did turn the stove on, and caused my neighbor's death by doing so, (a) was true at the time at which you said (2).

A modest assumption, one would surely think. One who makes it is not committed to supposing that (a) was highly probable at the time at which you said (2); much less that (a) was certain in the sense of having probability 1. I shall have nothing at all to say about what probability is in these remarks; but I shall take it throughout that our assuming that (a) was true at the time at which you said (2) is compatible with your having been right to think at that time that (a) was highly improbable, and indeed with (a)'s having *been* highly improbable.

But aren't we now on the road to Moore's view of the case? For shouldn't we agree that if a person will in fact cause his neighbor's death by turning his stove on, then he ought not turn his stove on—unless, of course, the circumstances are such that it is permissible for him then to cause his neighbor's death? So shouldn't we conclude, with Moore, that though you were entirely justified in saying (2), you spoke falsely in saying it?

I think it pays to spell this reasoning out in detail. What lies behind it is a very plausible-looking principle, viz.,

(IP$_1$) If A ought not cause B's death, then if it is the case that if A verb phrases, he will thereby cause B's death, then A ought not verb phrase.

I call the principle "(IP$_1$)" since it is the first of three "inheritance principles" I shall draw attention to; "inheritance principle" since it says that verb phrasing inherits impermissibility from causing B's death. I do think this a plausible-looking principle. If you may not cause a man's death, then surely you may not do that which you would cause his death *by* doing. Here is another sample inheritance principle:

(IP$_2$) If A ought not kill B, then if it is the case that if A verb phrases, he will thereby kill B, then A ought not verb phrase.

Also plausible. If you may not kill a man, then surely you may not do that which you would kill him by doing. So far so good. Now we were supposing that if anyone had said to me just before I turned the stove on,

(1) You ought not cause your neighbor's death,

he would have spoken truly. The modest assumption tells us that the outcome of my turning the stove on shows that

(a) If she turns her stove on, she will thereby cause her neighbor's death

was then true. So if (IP$_1$) is true, then you spoke falsely when you said

(2) It is permissible for you to turn your stove on.

But perhaps the fact that we have reached this conclusion will not seem objectionable; perhaps it now seems right to take Moore's line.

That was a case in which a person imposes a low, indeed an utterly trivial, risk of harm on another. What of cases in which a person imposes a high risk of harm on another?

What counts as a low or high risk of this or that harm is presumably a function, not merely of the probability of the harm, but also of the nature of the harm. (Just as the question in what conditions it is all right to cause a harm turns not merely on the conditions, but on the nature of the harm.) Thus a low risk of a bruise might well be a high risk of death. I simply sidestep the hard question, "What risk of what harm is low or high?"

A one-in-six chance of death on the spot is on any view a high risk of death. So let us imagine A is about to play Russian roulette on B—six chambers, one bullet. For a reason which will come out, I am going to suppose that the specially made 'roulette gun' with which A is about to play the game works like this: you aim the gun, you press a button on the handle, that starts the cylinder spinning, and if there is a bullet under the firing pin when the cylinder stops spinning, it is fired in the direction of aim. B, let us suppose, is asleep; he is no threat to anyone, and there is no great good which might be accomplished by his death. So we say to A:

(3) You ought not press that button.

Let us suppose that A cares nothing for that and proceeds to press the button. The cylinder spins; and when it stops spinning, the bullet is not under the firing pin, so the bullet is not fired. B is not killed; he is in no way harmed; let us suppose he is not even awakened by the small click which the cylinder made when it stopped spinning. Does that show that we spoke falsely when we said (3)?

One is surely inclined to say no. I said in section 4 above:

> Suppose A played Russian roulette on B. B has ground for complaint against A even if B was caused no harm (no bullet was under the firing pin when A fired), and even if B was unaware of what happened so that he was caused no fear or discomfort. The ground for complaint lies in the fact that A imposed a risk of death on B.

Surely it was true to say that A ought not press the button despite the fact that no harm of any kind came to B.

What if, not only did no harm come of A's pressing the button, but some good came of it? It is not easy to imagine a good which might have come of it, but suppose a small good did come of it. Even so, one is surely inclined to say that that makes no difference. Even so, we spoke truly when we said to A that he ought not press that button.

Moore is not in fact committed to saying that our inclination to say this is a mistake. Moore's utilitarianism commits him to saying that it was permissible for A to press the button only if there was no other course of action open to A at the time which would have had better consequences than those which his pressing the button actually did and will have. And it might well have been the case that there were courses of action open to A at the time which would have had better consequences. But if we suppose (no doubt per impossibile) that the only other course of action open to A at the time was standing still, with eyes closed, and counting to ten by twos, then on the usual assumption about what engaging in that course of action causes, it was permissible for A to press the button; and if A's pressing the button caused some good, however small, then pressing the button was something A quite positively ought to do.

Is there anything to be said for taking Moore's line here? Not much, no doubt, but not nothing.

Why did A press the button on his roulette gun? People who play Russian roulette on others presumably do so in order to get the fun of imposing a high risk of death on those others; and let us suppose that this was the point in it for A. A bad intention, if there ever was one. And so Moore would say: that shows that although A's pressing the button was not impermissible, and perhaps was even something he ought to do, he is all the same to blame for doing it.

As I implied earlier, there is something more that could be said although Moore does not in fact say it, namely, that A's act of pressing the button was a morally bad act. (Compare the possibility of saying that, not merely am I not to blame for my neighbor's death, but also that my act of turning the stove on was not a morally bad act.)

And it pays to notice that there is a further possibility. It could be said in addition that—the outcome of the button pressing having been what it was—although we spoke falsely when we said

(3) You ought not press that button,

we would have been speaking truly if we had said

(4) You ought not press that button in order to get the fun of imposing a high risk of death on B.

I think that the idea that the falsehood of (3) is compatible with the truth of (4) is one which we have good reason to want to make room for in any case. The point in fact is a familiar one. Here is a more familiar kind of example. Jones has a certain stuff which in fact is a medicine, which Smith needs for life. Jones, however, thinks the stuff is a deadly poison, and in that he wants Smith dead, if he gives the stuff to Smith he will give it to Smith only in order to cause Smith's death. The fact that he has that belief and desire surely does not make

(3') Jones ought not give the stuff to Smith

true. Indeed, other things being equal, (3') is false. (Given, for example, that it is not the case that Jones stole the stuff from Dickenson, who also needs it for life. Given that Jones has not promised to give it to Bloggs, whose child needs it for life. Etc.) What is true is, rather,

(4') Jones ought not give the stuff to Smith to cause Smith's death.

(4') is plausibly viewable as true no matter what the consequences of Jones's doing what *it* says he ought not do may turn out to be; and the truth of (4') is surely compatible with the falsehood of (3'). As I said, the point is a familiar one: it is that "ought" attaches, not to acts, but to activities or act kinds. So one who says that we spoke falsely when we said (3), and would have spoken truly if we had said (4), is not thereby convicted of inconsistency.

So can it be said that the reason why we feel that, whatever the outcome, we spoke truly when we said (3) is that we are swamped by the thought that A was to blame for acting as he did, and by the fact that his act of pressing the button was a bad one, and by the fact that there was something he did which he ought not have done, viz., press the button in order to get the fun of imposing a high risk of death on B? And, in addition, the fact that we were so patently justified in saying (3) to him at the time?

Compare Bloggs. Bloggs is in his kitchen, about to press the button on his electric toaster. Why? To make toast, of course; that is his only aim, as it always is, every morning. We are a representative from Con Ed, examining the wiring in the basement. What a weird thing we discover! The wire from the circuit which Bloggs's toaster is on has been attached (by some villain, no doubt) to a roulette gun trained on Dickenson (Dickenson is Bloggs's neighbor, and he sleeps late). We come tearing up the back stairs, knowing of the breakfast Bloggs always eats; we call out, "You ought not press that button!"

Notice, in passing, that we do not call out, "It is perfectly all right for you to press that button *now*; but we have some information such that after we have given it to you, it will *then* be the case that you ought not press that button." We do not suppose that what a man ought or ought not do turns on what he thinks is the case.

Alas, too late. Bloggs has already pressed the button. Fortunately all went well for Dickenson; no harm came to him. Did we speak truly when we said (3), despite the happy outcome of the episode? "Thank goodness," we say; "what luck!" Should we be taken to have in mind "What luck; although you did something you ought not have done, no harm came of your doing it"? Or should we instead be taken to have in mind

"What luck; as things turned out, you did nothing you ought not have done"? Is it supposed to be obvious that the former is the preferable account? I do not think that is obvious.

What comes out, I think, is that A's belief and bad intention do play a role in generating our thought that we spoke truly when we said (3) to A. For in the case of Bloggs—who differs from A only in belief and intention—it is not obvious that we spoke truly when we said (3) to him. And is it right to assign belief and intention that role? The case of Jones suggests it is not. For the fact that if Jones gives Smith the stuff he will give Smith the stuff in order to cause Smith's death should surely not be taken to warrant saying that Jones ought not give Smith the stuff.

Perhaps it would be helpful at this point to bring these cases together with the case of me and my stove, and list the options.

(I) We can take Moore's line on all of them. Thus we can say the unhappy outcome of my turning the stove on shows that you spoke falsely when you said

(2) It is permissible for you to turn your stove on,

and that the happy outcome in the cases of A and Bloggs shows that we spoke falsely when we said to them

(3) You ought not press that button.

I have been defending Moore's line because I am inclined to think that people have not sufficiently appreciated what can be said for it, and how difficult finding a plausible alternative turns out to be.

On the other hand, I do think we had better try. It *is* counterintuitive to say that it was not permissible for me to turn on the stove, and that it was permissible for A to press the button. It is much less counterintuitive to say that it was permissible for Bloggs to press the button; but perhaps we really should agree that it was true to say to him, as we did, that he ought not press it.

People sometimes wonder whether it is all right for them to take a chance; and we often feel it was, even if things turn out badly. Here are five, who are starving. They can be saved if and only if Jiggs saves them; but Jiggs will save them if and only if I cause Smith's death by dropping a very heavy weight on his head. Most people are inclined to think I may not do so. But now here are another five, who are also starving. They can be saved if and only if I fly out in my plane and drop them a food parcel. If I do drop the parcel, I impose a risk of death on Smith, for Smith is out in a nearby meadow, tending his sheep, and a sudden gust of wind just might, conceivably, blow the falling food parcel onto Smith's head. Is it permissible for me to take that chance? Of course. The risk is utterly trivial. And I think we do feel inclined to go on saying it was permissible for me to take that chance even if, in the event, a sudden gust of wind does blow the food parcel onto Smith's head. We do not think that the permissibility of acting under uncertainty is settleable only later, when uncertainty has yielded to certainty.

If these ideas are right—and it really does seem that they are—then risk imposition does generate an independent problem for moral theory. For there is a further question which then arises, beyond the question what harms we may or may not cause in what circmstances, viz., the question what risks of what harms we may or may not impose in what circumstances.

So we need an alternative to Moore's line. And what should it be?

Earlier I spelled out a bit of reasoning which issues in Moore's conclusion about the case of me and my stove. It issued, that is, in the conclusion that you spoke falsely when you said

(2) It is permissible for you to turn your stove on.

Which of the premises of that bit of reasoning should be given up?

(II) I mention first, only to dismiss it, the possibility that we should give up the modest assumption, and in particular, the assumption that the outcome of my turning the stove on shows that

(a) If she turns her stove on, she will thereby cause her neighbor's death

was true at the time at which you said (2). Giving that up strikes me as a dismal idea, and I suggest we pass it by.

(III) A more interesting idea is to give up the inheritance principle

(IP_1) If A ought not cause B's death, then if it is the case that if A verb phrases, he will thereby cause B's death, then A ought not verb phrase.

Abstract principles in moral theory are always suspect; perhaps it was wrong to think this one so plausible? Perhaps we should say that although it was then true that I ought not cause my neighbor's death, it was nevertheless permissible for me to turn the stove on—despite the fact that if I turned it on I would cause my neighbor's death by turning it on?

But one can't just give (IP_1) up and wash one's hands of the question what light (or dark) doing so sheds on the rest of moral theory. One can't just say: I get out of my trouble if I reject (IP_1), so I reject (IP_1). For example, are we to say that all inheritance principles of the same form as (IP_1)—viz.,

If A ought not VP_1, then if it is the case that if A VP_2s, he will thereby VP_1, then A ought not VP_2—

are false? Or only some subset of them? And if we are to say that all of them are false, are there any inheritance principles of any other form which are true? Is there no "nesting" of impermissibilities? Does every set of ascriptions of impermissibility such that no one of them entails any other form a mere concatenation, each requiring independent justification?

(IV) I think it pays to take note of the remaining possibility—though it is probably no more than a mere possibility—namely that we should retain (IP_1), and instead say that you would have spoken falsely if you had then said

(1) You ought not cause your neighbor's death.

It sounds a very odd idea, in light of the fact that there was nothing in the circumstances in which I was then placed which would have justified my causing my neighbor's death; but I think it worth taking note of.

Suppose we were to say, quite generally, that it just never is true to say of anyone, whatever the circumstances, that he ought not cause another person's death. Suppose we were to say that what is true is at most that a person ought not impose a high risk of death on another.

Consider the following inheritance principle:

(IP_3) If A ought not impose a high risk of death on B, then if it is the case that if A verb phrases, he will thereby impose a high risk of death on B, then A ought not verb phrase.

This principle will not generate an argument to the conclusion that you spoke falsely when you said it was permissible for me to turn the stove on; for it was not true that if I turned the stove on, I would impose a high risk of death on my neighbor by doing so. I have been supposing that you were right to think it highly improbable that I would cause his death by turning the stove on. By contrast, it will generate an argument to the conclusion that we spoke truly when we said to A—and to Bloggs—"You ought not press that button." For of both it was true that they would impose a high risk of death on a person by pressing the button.

Notice that while taking this line requires giving up some of the "objectivity" of "ought," it does not require that all be given up. For if we take this line, we are not committed to supposing that what counts is what the agent thinks on the matter. Thus, in particular, it allows us to say that it not only was permissible for me to turn the stove on, it would have been permissible for me to do so even if I had thought (falsely) that doing so would impose a high risk of death on my neighbor. For my doing so would not in fact have imposed a high risk of death on him. I should think that at least that degree of objectivity must be retained whatever line we take on these matters.

But perhaps it really is too counterintuitive to say that it just never is true to say of anyone, whatever the circumstances, that he ought not cause a person's death. Moreover, I think that if we took this line, we should also have to say that it never is true to say of anyone, whatever the circumstances, that he ought not kill a person; and that is, if possible, even more counterintuitive. But I think that we would have to say this further thing. For didn't I kill my neighbor by turning my stove on? (Given that my turning the stove on caused an explosion in his apartment, and he died because he was in the apartment at the time.) Presumably we could still have it that a person ought not shoot a man in the head, or strangle him, and so on, for doing these things imposes a high risk of death on the victim. But we should be unable to say that a person ought not actually kill another; and that sounds quite unacceptable.

So option (III) is probably the best. But I leave the matter open, not having any clear idea how to answer the questions I raised in discussing it.

6. I have been assuming throughout that B does not consent to A's imposing the risk of harm on him, for if he does consent to this—if he wittingly and freely consents to it—then it seems plausible to think that no problem arises: A may impose the risk.

There are notorious difficulties which arise about consent; but there is one which I think has not been much attended to. I do no more than draw attention to it here.

Richard Posner (for purposes which are not relevant here) said the following in a recent article:

> I contend, I hope uncontroversially, that if you buy a lottery ticket and lose the lottery, then, so long as there is no question of fraud or duress, you have consented to the loss.[3]

Vain hope! His contention was objected to, indeed sneered at, by Jules Coleman and Ronald Dworkin in their commentaries on Posner's article.[4] But what exactly is their objection? I think that what Dworkin has in mind is the same as what Coleman has in mind in writing as follows:

> If I buy the lottery ticket and lose my loss may be a fair or legitimate one, one that it may be appropriate to pin on me. It would be fair because I had willingly taken a risk by consenting to or voluntarily joining an enterprise that was risky in the relevant way. But it would hardly follow that I had consented to the loss.[5]

And what does Coleman have in mind in writing those sentences? I hazard a guess it is the following. Suppose that A offers B an \$n ticket in a lottery A is running. B buys it. Then the following is true:

(1) B has consented to its being the case that (If A does not draw B's name on ticket-drawing day, then B loses his \$n).

Alas for B, the following turns out to be true:

(2) *A* does not draw *B*'s name on ticket-drawing day.

I hazard a guess that Coleman's and Dworkin's point is this: it does not follow that

(3) *B* has consented to its being the case that *B* loses his $n

is true. I.e., the conjunction of (1) and (2) does not entail (3).

If that is their point, then I should think they are right. Modal principles of the form

$$[M(p \supset q) \ \& \ p] \supset Mq$$

have always to be watched; and the moral principle which results from writing in "*B* has consented to its being the case that" for "M" really does look suspect.

But if that is their point, then why the sneering? What if Posner had instead written:

> I contend, I hope uncontroversially, that if you buy a lottery ticket and lose the lottery, then, so long as there is no question of fraud or duress, you have no ground for complaint about your losing.

That contention looks much more plausible. And so far as I can see, nothing Dworkin says makes any trouble for it at all. (Of course, whether saying that would have served Posner's purposes as well as saying what he did in fact say is quite another matter. I think it would have. But the question whether it would have has no bearing on our present concerns.)

But there is something which Coleman says which makes trouble for the hope of obtaining an easy generalization from it; Coleman has a further point (I do not think he sees it is a further point) and it is that which I think raises the interesting difficulty. What I have in mind comes out in the following example, which was suggested by Coleman's example.[6] Suppose there are two ways in which I can get home from the station at the end of the day. The first is pleasant, passes through a brightly lit middle-class shopping area, is quite safe, but is long. The second way is unpleasant, passes through an ill-lit area of warehouses, is unsafe, but is short. Nobody has ever been mugged while walking along Pleasant Way; people have from time to time been mugged on Unpleasant Way. Here I am, at the station; I am tired; I think, "The hell, I'll chance it, I'll take Unpleasant Way." I then promptly get mugged. I surely have grounds for complaint — at least against the mugger, and perhaps also against the city.

Perhaps it was foolish to walk through Unpleasant Way at night. (But the more tired I was and the less risky Unpleasant Way is, the less foolish it was.) And my friends can later say, "Well, it would be out of place for you to feel *surprised* at the fact that you got mugged." But that is different from "Well, it would be out of place for you to feel *wronged*." That comment would itself be out of place.

But didn't I consent to the risk of being mugged? And there was no fraud, no misperception of any kind. I knew perfectly well what the risk was. And there was no duress, no coercion or compulsion. I was not under pressure from any person to choose that route home. Of course I chose that route because I was tired. But my fatigue no more coerced or compelled me than any person did.

It is plausible to think that one who loses in a nonfraudulent lottery, which he entered without duress, has no ground for complaint when he loses. Why is it that I have ground for complaint when I get mugged on Unpleasant Way? I think it a nice problem.

I think that finding an answer requires looking into the content of the consent. More generally, that it is important not merely to look at whether a person's consent is witting and free (that is where most of the attention has been focused), but also at what exactly he consented to. (We talk altogether too glibly of "consenting to a risk.") It was easy enough to say what *B* had consented to when he accepted *A*'s offer of a lottery ticket (cf. (1) above). What did *I* consent to in the story I just told?

In the first place, and obviously, there plainly is no person such that I consented to his mugging me.

Second, there is no person or persons such that I consented to his or their imposing a risk of being mugged on me. (In any case, nobody did impose that risk on me. Compare Unpleasant Way with a part of the city in which people indulge in playing Russian roulette on passersby. If I had walked through there, and somebody had played Russian roulette on me, somebody would have imposed a risk on me. But not by my consent.)

Third, I did not consent to Unpleasant Way's being dangerous to walk through at night, or to its being the case that one who walks through it at night is at risk of being mugged. So in particular, I did not consent to its being the case that if I walk through Unpleasant Way at night, then I get mugged; or even to its being the case that if I walk through Unpleasant Way at night, then I am at risk of being mugged.

I am inclined to think that the most that can be said is this: knowing that Unpleasant Way is dangerous to walk through at night, knowing that one who walks through it at night is at risk of being mugged, I consented to its being the case that I walk through Unpleasant Way at night. (Compare the workman in a risky factory. I am inclined to think that the most that can be said of him is this: knowing the factory is risky, he consents to work in it.) And I am inclined to think also that if that is the most that can be said, then my consent goes no way at all toward removing my ground for complaint about what happened to me on Unpleasant Way.

But those ideas call for more, and closer, attention than I can give them now; and it may be that there is no future in them in any case.

Notes

I am indebted to all of the participants in the Working Group for their criticisms of an earlier draft of the present chapter, but in particular to Mary Gibson and Douglas MacLean.

1. In Cambridge recently, a man brought charges against his neighbor on ground of public nuisance: the neighbor had not mowed his lawn in fourteen years. Plaintiff lost.
2. G. E. Moore, *Ethics* (London: Oxford University Press, 1912), p. 121.
3. Richard Posner, "The Ethical and Political Basis of the Efficiency Norm in Common Law Adjudication," *Hofstra Law Review* 8, no. 3 (Spring 1980): 492.
4. Jules Coleman, "Efficiency, Utility, and Wealth Maximization," and Ronald Dworkin, "Why Efficiency?", in *Hofstra Law Review*, Note 3 above.
5. Coleman, "Efficiency," pp. 534–35.
6. Ibid., fn 45, pp. 536–37.

7

Consent and Autonomy

Mary Gibson

As we have seen, air pollution poses risks to human health in a variety of ways, ranging from secondhand cigarette smoke to workplace toxins to outdoor ambient and long-range transportable pollutants. The feasibility of obtaining individual consent from those at risk in these different contexts varies widely. How can public policy concerning air pollution give appropriate weight to the values that make individual consent important while facing up to the fact that individual consent is not possible in many situations where policy decisions must be made? If we can identify what it is that makes consent morally important, we may be able, in situations where consent is not feasible, to obtain some guidance by exploring the implications of the values underlying the moral significance of consent.

In Chapter 4, Samuel Scheffler presents three different conceptions of the moral importance of consent. In this chapter, I propose to explore the implications of what I take to be a version of one of those conceptions to see what guidance it offers for the three specific arenas we have chosen to focus on: secondhand cigarette smoke, workplace air pollution, and outdoor air pollution. In particular, I suggest that the ideal of autonomy, understood as self-determination or being in control of one's life, is the central value behind the moral significance of consent. First I shall discuss the view I favor concerning the moral significance of consent. Then I shall consider each of the three arenas in turn, inquiring how far we can get with straightforward consent, and what guidance, if any, the underlying view provides for dealing with the remainder.

Why Do We Care About Consent?

The first of the three views Scheffler outlines concerning the importance of consent is that it has instrumental value: consent requirements may help to deter conduct that is morally unacceptable for reasons having nothing further to do with consent; they may enhance self-esteem and feelings of personal responsibility, promote solidarity, and reduce alienation and hostility.

The second view is that consent has intrinsic moral value: it is an essential component of a good life. Whatever else it involves, a good life must include opportunities for individual exercise of capacities for judgment, decision, and commitment, and some degree of control over the allocation of time and energy to the different components of that life. This is "not just because one is more likely to end up with satisfying pursuits if one has some say in their selection. The process of selecting has an added value of its own."[1]

On the third view, the importance of consent derives from its role within a system of natural rights. Rights, on this view, are morally fundamental and may be overridden, if at all, only to avert catastrophe. Consent provides a way for individuals to interact and to coordinate their activities without violating each other's rights.

As Scheffler points out, these views are not mutually exclusive. While a utilitarian might hold the first view exclusively, others might combine the first and second views with various degrees of emphasis on the instrumental and intrinsic value components. (Such a combination, indeed, is the sort of conception I propose to explore.) A proponent of the third view might maintain that one or both of the first two conceptions raises considerations that, while secondary to respect for rights, do contribute to the importance of consent. Similarly, proponents of the first two need not be unsympathetic to talk of rights, although they are likely to differ from adherents of the third view by allowing a greater role for consequences in determining both what rights there are and when they may be overridden.

It is not entirely clear why it matters, not just that one's life be happy, however that may be understood, but that it be, so to speak, one's own. But it does matter. In something like the same way, it matters that I am not just a brain in a vat being stimulated by some scientist so that it seems to "me" that I am a complete human being sitting in a slightly chilly room writing, while rain splashes on the windows. I would not trade *my* life for one determined by another, regardless of whether that other was a mad brain surgeon or a wise and benevolent decision maker who intervened only to prevent me from making decisions that would be likely to decrease my happiness. Even if I were reliably guaranteed that the new life would be happier, overall, than my own life, I would not trade. We do not value autonomy simply, or even primarily, because we think it makes us—or people generally—happy.

Of course, the cases of the mad brain surgeon and the wise but meddling decision maker are not entirely parallel. The brain-in-a-vat story typically comes up in discussions of skepticism, where the point of it is that, even though it matters very much to us, we cannot know for certain that we are not brains in vats because, by hypothesis, if we were, all our experiences would be—or seem to be—just as they are now. We often *can* tell, on the other hand, when someone intervenes in our lives to make for us decisions that otherwise we should have made for ourselves.

Yet sometimes we *cannot* tell. Often, the most effective forms of intervention are those that leave us believing either that the decisions were in fact our own or that there were no available alternatives and hence no decisions to be made. We raise no objections to such interferences, but only because we are not aware of them. For that very reason, we tend to regard such covert intervention as even more objectionable than straightforward overt ways of overriding a person's autonomy. Our reaction is, not that what we don't know can't hurt us, but rather that we cannot fight an enemy we cannot see. The value we place on autonomy cannot be reduced to the satisfaction we experience in seeming to be making our own decisions, living our own lives.

Autonomy—self-determination—has for us intrinsic value. That is why, even in circumstances where the potential negative consequences of a person's own choice clearly outweigh the potential benefits to that individual and to others, it often is not justifiable to override that person's choice. The intrinsic value of the exercise of autonomy in the making and implementing of that choice must be placed in the balance. And if autonomy is intrinsically valuable, then its exercise must be so as well, for there can be autonomy only in and through its exercise.

Why we value it is not the only puzzling thing about autonomy. It is also difficult to say just what it is, and hence whether it is a realizable or even a coherent ideal.

A crucial presupposition of the ideal of autonomy is that beings capable of it are capable of choice. One may choose more or less consciously, more or less deliberately or spontaneously, and more or less autonomously. But for these ideas to have any application at all, there must be such a thing as genuine choice. It must not be that people merely appear to themselves and each other to be choosers. But it has seemed to some people that, if all of our actions are causally determined, there can be no such thing as genuine choice. This is not the place for a discussion of the free-will vs. determinism controversy. For present purposes, I shall assume that determinism implies, not that we are not genuine choosers, but that our choices are determined by factors over which we have various degrees of control, including many over which we have little or no control. I am assuming, in other words, that the truth of determinism would not automatically rule out the possibility of autonomy by eliminating choice.

It is worth stressing, though, that choice is merely a presupposition of autonomy. As R. S. Peters says,

> Our normal expectation of a person is that he is a chooser—that he can be deterred by thoughts of the consequences of his actions, that he is not paranoid or compulsive and so on. But such a person might be a time-serving, congenial conformist, or an easy-going, weak-willed opportunist. Being a chooser is a standard expected of anyone . . . it is not an ideal of conduct or of education. . . . What, then, has to be added for a chooser to develop into an ideal type of character in which being free features? To ask this is to ask for the criteria for calling a person 'autonomous.'[2]

I have spoken of autonomy in terms of self-determination, of being in control of one's life, and of one's life being one's own. As its etymology suggests, to be autonomous is to be governed by laws or rules one adopts for oneself. But this skeletal idea can be fleshed out in various ways, some of which seem to me clearly untenable, and none of which is obviously free of difficulties. In the remainder of this section, I shall be attempting to point in the direction that I think an adequate account of autonomy must go. A fully adequate account, if one can be developed, would show autonomy to be a coherent, realistic ideal that is worthy of being valued. I cannot hope fully to accomplish that task here.[3]

What Autonomy Is Not

According to Immanuel Kant, the law that an autonomous individual gives to him or herself is derived from the individual's nature as a rational agent alone and must be independent of the individual's empirical circumstances, both social and natural, such as personal history, social environment and relations, values and commitments, and physical and emotional needs and desires.[4] But I am unable to make sense of this idea. I do not understand how a purely rational agent could come to any decisions at all, independently of all empirical circumstances. I not only cannot imagine what would prompt such an agent to decide one way rather than another; I cannot imagine what would prompt it (sex and gender are empirical circumstances)[5] to decide at all. Since this idea is unintelligible to me, autonomy cannot, for my purpose, be distinguished from heteronomy (rule from outside oneself) in the Kantian way. But then how are we to distinguish autonomy from heteronomy?

An autonomous person acts for reasons that are, in some important sense, his or her "own," and not for reasons (or nonreasons) that come, in some important sense, "from outside." But what are these senses in which reasons for action can be either one's "own" or "from outside"? Surely, in the most straightforward sense, most, if not all, reasons come largely, if not entirely, from outside. If you offer a convincing argument that my doing A would be better, all things considered, than my doing B, and, accepting your argument, I do A, I have not thereby failed to act autonomously. If I start out to drive 50 miles to visit you but turn back because of an impending storm, I do not thereby fail to act autonomously. Yet both of these reasons come, in the most obvious and literal sense, from outside—either from another person or from the rest of the natural world. That your argument and the impending storm constituted reasons for me, of course, depended upon certain features of me, my values, desires, beliefs, and so forth. But these, in turn, came largely or entirely from outside, too, in the process of my informal and formal socialization and education. Surely, whatever there may be in our natures that may be termed innate, it will not suffice to give us a reason for acting one way rather than another in very many of the situations in which we actually do one thing rather than another.

If autonomy were the expression of some innate nature, it would be a given, a fact of human life, to be valued and respected perhaps, but not an ideal to strive for in our lives and to nurture and promote in our relations with others and in our social institutions, policies, and practices. Respecting autonomy, in that case, would amount at most to refraining from imposing external obstacles in the way of its expression. There would be no sense to the idea that one could fail on one's own to think and act autonomously, no basis for concern that certain kinds of individual or social relations and practices might prevent or hinder the attainment of autonomy by those involved in them.

If autonomy is to be a coherent and realistic ideal, for reasons or rules to be "one's own" cannot mean that they literally originate inside oneself. Rather, whatever their source, one makes them one's own in the sense required for autonomy, by critically examining and rationally reflecting on them, and by committing oneself to and acting upon them.

Autonomy as a Social Product

Psychologist Jean Piaget employs a conception of autonomy that seems to me very promising.[6] Autonomy, on this conception, is not a fundamentally individualistic idea that must somehow be accommodated to the undeniable interdependency of individuals with one another and their environment. Rather, autonomy is, in its origins and at its very core, a social phenomenon resulting from relations of cooperation and attitudes of mutual respect among persons who regard themselves and each other as equals.[7]

Piaget distinguishes between two kinds of social relations: relations of constraint and relations of cooperation. Relations of constraint are compatible with, and even encourage, heteronomy in those who are the objects of constraint. Rules and orders are imposed on them by an external authority and are enforced by the threat of punishment together with the sheer weight of authority. The required behavior may have no relation to the purposes or desires of those on whom the requirements fall. The main reason for complying is to avoid the sanctions, i.e., punishment or the displeasure of the authority. A compliant, unquestioning attitude is fostered, not only toward rules for behavior, but toward beliefs, ideals, values, tastes, and pursuits. Individuals are encouraged to accept as "given" the beliefs, rules, values, etc. that are handed down from on high, so to speak. The individual may continue to regard them as "external" and conform in order to avoid making waves (having no alternatives to compare them with, test them against, etc., there is no reason not to conform). Or the individual may swallow whole the prevailing views and values, so to speak, without chewing and digesting them: without critically examining and rationally reflecting on them, and thus making them his or her own. (In some cases, as in very young children, where the individual does not clearly distinguish between self and not-self, the difference between these two attitudes will be blurry or nonexistent.)

Relations of cooperation, on the other hand, encourage autonomy. Rules of behavior are seen as necessary conditions for carrying out the cooperative enterprise. They are not imposed from above, but are mutually agreed upon by those participating in the cooperative activity. They are subject to discussion and criticism by any participant and may be changed by mutual consent. The rules one accepts and acts upon are directly related to one's own pursuits and purposes and are seen as necessary for advancing those pursuits and purposes. They are not viewed as fixed or final, but are to be questioned, examined, criticized, and, when warranted, revised. Thus, a critical, questioning posture is encouraged of rational reflection on what is done and why.

Piaget's work investigating the intellectual and moral development of children indicates that two important features of the lives of young children tend to reinforce each other. First, inegalitarian relations of constraint predominate in their lives, the young child being subordinated to and constrained by parents, other adults, and older children. The second feature is childish egocentrism, or the inability of the child to distinguish clearly between self and not-self.[8]

As the child gets older, however, egalitarian relations with peers have an increasing role in his or her life. The child develops a sense of self by interacting with and learning to recognize other selves.

> Now, insofar as constraint is replaced by cooperation, the child disassociates his ego from the thought of other people. For as the child grows up, the pres-

tige of older children diminishes, he can discuss matters more and more as an equal and has increasing opportunities . . . of freely contrasting his point of view with that of others. Henceforward, he will not only discover the boundaries that separate his self from the other person, but will learn to understand the other person and be understood by him.[9]

Critical thinking and rational reflection, capacities essential to autonomy, develop together with—making possible and made possible by—cooperative social relations.

> Cooperation alone leads to autonomy. With regard to logic, cooperation is at first a source of criticism; thanks to the mutual control which it introduces, it suppresses both the spontaneous conviction that characterizes egocentrism and the blind faith in adult authority. Thus discussion gives rise to reflection and objective verification. But through this very fact cooperation becomes the source of constructive values. It leads to the recognition of the principles of formal logic insofar as these normative laws are necessary to common search for truth.[10]

Thus, relations of cooperation give rise to such intellectual norms as logical thought and expression and intersubjective verification, as well as to such moral norms as reciprocity and mutual respect. The "inner dialogue" in which an autonomous person critically examines and rationally reflects on her or his views and values imitates and is made possible by participation in egalitarian social relations.

On the basis of what has been said so far, we can begin to fill out the notion of autonomy as one's life being one's own. First, one's choices must be genuine choices, not illusory. Second, they must result from critical examination of the available alternatives and rational reflection on the potential outcomes.[11] Third, this examination and reflection must be done in the light of beliefs, values, goals, tastes, etc., that are one's own in the sense that they are neither conformed to for convenience (and lack of a conception of alternatives) nor swallowed whole.

If one's life being one's own is to include self-determination in the sense of effective control of one's life, however, then there must be more to autonomy than this. We might speak of two aspects of autonomy: internal capacities of critical examination and rational reflection involved in shaping oneself, so to speak, and external conditions involving the absence of obstacles to one's actively shaping one's environment in accordance with one's conception of oneself and one's goals. But let us pause to reflect on this way of speaking.

The "Internal-External" Distinction and the Bounds of the Self

I am not entirely comfortable with the language of internal and external aspects of autonomy because it tends to encourage us to think in terms of two distinct realms that interact in somewhat mysterious ways, but each of which has its own independent nature, so to speak. In contrast, I believe we should think in terms, not just of interaction, but of interdependence and interpenetration profound enough to make the notion of two distinct realms seem inappropriate (and also to make the interaction seem less mysterious, though still far from adequately understood).

Does this raise problems for the idea of autonomy as self-determination by blurring or even obliterating the boundaries of the self? Perhaps it does, but, if so, then so much the worse for the idea of autonomy. For it seems to me that no adequate conception of autonomy can take the internal-external distinction as determining the boundaries of the self. Indeed, it seems to me that we shall need a conception of the self that permits of flexible boundaries if we are to make sense of autonomy as self-determination. The boundary will sometimes have to divide some internal features from others, as when one must control, alter, or even eliminate some aspect of one's character, personality, or psychological makeup if one is to be the sort of person one wants to be, to shape oneself and one's life in the way one wants. Moreover, this internal division cannot be a fixed one, marking off some core self from the rest in such a way that if (or to the extent that) the core self rules the whole self, then (or to that extent) the individual is autonomous, or meets the internal conditions for autonomy. One candidate for the core self, i.e., the purely rational self, has already been rejected for this role because it leads to an incomprehensible account of autonomy. That no other fixed line will do seems to follow from the claim that, although some beliefs, values, personality traits, etc. are more central to what we are and want to be than others, none is in principle beyond question, criticism, and revision. To be able to subject one's most central beliefs and values to such scrutiny, one must be able to distance oneself from them to some extent (though not all at once, of course). And though it may in fact seem inconceivable, it must be logically possible that one could, upon examination, reject any one of them.

Extending the bounds of the self outward, beyond the inner-outer line, may also be appropriate for some purposes. In the same way that one makes beliefs and values part of oneself by identifying oneself with and committing oneself to them, one can make another person, a family, a group, a community, etc., part of oneself by identifying with and committing oneself to them. This need not mean either that the group becomes submerged in the individuals who identify with it or that the individuals, by so extending the boundaries of their selves, lose their individuality by submerging themselves into the group. When one adopts a new value, goal, or belief, one does not, in doing so, abandon all one's previously held views and values and become indistinguishable from others with whom one shares the new part of oneself. One may, of course, be introducing a new source of tension into one's set of beliefs and goals. For even if one avoids introducing (or succeeds in eliminating) any outright contradictions by rejecting any previously held views and values that may be logically incompatible with the new, it is never guaranteed that situations won't arise in which some of one's values or beliefs come into conflict with others. Indeed, this happens to everyone every day in the most mundane ways, and occasionally to most people in much more dramatic ways. In some cases, one decides that value A is weightier than value B and should always win out when they come into conflict.[12] In another case, value A may win out, given the circumstances, while not being judged weightier than B in general. Some such conflicts may be resolved by flipping a coin or alternating between choices that favor A and those that favor B. In such cases, it is possible to have tensions and conflicts among one's values, beliefs, etc., and to resolve the conflicts, permanently or provisionally, in one's choices and actions while continuing to hold all of the competing values and views. The choices may be difficult and painful, and one may be unsure that one has chosen well, but one may not be forced to abandon or repudiate any of one's commitments.

In other cases, however, one may find that, although there is no logical contradiction among them, one cannot in one's actual life circumstances pursue all of one's goals or promote all of one's values. Some of them, even some very important and central ones, sometimes must be abandoned. In such a case, one is likely to have a serious sense of loss, of giving up something of oneself, of being less than one was. To make matters worse, one may be plagued by a nagging uncertainty about whether one has made the right decision. (Perhaps partly for this reason, previously held views and goals are sometimes repudiated with particular vehemence. This seems frequently to be the case with political positions, for example.) In some such cases, one is unable to resolve the conflict, even provisionally, and is virtually paralyzed by it—unable to pursue any of one's goals effectively.

Similarly, then, when one extends the boundaries of oneself through commitments of love or solidarity to other persons or groups, one does not thereby lose, abandon, or submerge all of one's previously held commitments, views, and values. Honoring the new commitment is very likely to involve overriding some of one's previous preferences and pursuits—else why call it a commitment? And it is likely on some occasions to come into conflict with other important, even central, values, goals, and commitments. There is no automatic resolution of such conflicts, either in favor of the group or in favor of the individual standing by his or her principles. Not in favor of the group because, although commitment to the group is one of the individual's constitutive features, by hypothesis, since there is a conflict, it is not the only one. Not in favor of the individual standing by his or her principles because, although it is not the only one, commitment to the group is one of those constitutive features: standing by the group is one of the principles the individual is committed to standing by.

Again, in some cases it will be possible to resolve the conflict one way or another without giving up any of the competing commitments, and in some cases it will not. If nothing is, in principle, beyond question, criticism, and revision, then even the deepest commitments do not, in cases of genuine conflict, provide automatic answers.[13]

For these reasons, then, I believe that an adequate account of autonomy will have to admit of a conception of the self with flexible boundaries. Such flexibility may help to explain why it is so difficult to pin down the self (one's own or another's), to determine once and for all what belongs and what does not. For some purposes, my self extends back to my birth or early childhood while for others it is of much more recent vintage; for some purposes it includes my love for my sister, the pain in my back, or my taste for spicy foods while for others it does not.[14]

Having said a bit about why I am uncomfortable with the internal and external terminology, and why I do not think the distinction it purports to draw is needed for an account of the self or of autonomy, I shall continue to use this terminology for lack of a better alternative. This should be harmless provided that one is careful not to take it as signifying two distinct realms.

Conditions for Autonomy

It is important to see, then, that autonomy as control of one's life involves both internal capacities and external conditions—social and physical. These capacities and conditions are causally interrelated. In addition, social relations, practices, and institutions profoundly affect not only the external conditions—in both obvious and not-so-obvious ways—but also the development and exercise of the internal capacities. If we are committed to autonomy, then, we must be prepared to examine the ways in which our social arrangements, processes, and policies are likely to influence, promote, or hinder the development and effective exercise of autonomy in the persons affected by them.

I said earlier that if autonomy is valuable, then its exercise must be so as well, for autonomy exists only in and through its exercise. As Piaget argues, persons develop the capacities for critical examination and rational reflection by engaging (initially with their peers, and then also on their own) in those same activities. The more one exercises these capacities, the further they develop, so that one is not only developing necessary conditions for autonomy, more and more one is, at the same time, exercising autonomy. The degree to which one exercises autonomy is the degree to which one is autonomous. The importance to autonomy of actually going through the process of deliberation, choice, and implementation has implications for the degree to which appeals to such notions as hypothetical or indirect consent may be justified.

For example, appeals to hypothetical consent, where it is maintained that the individual *would* consent *if* he or she were to go through the process, do not preserve the value of autonomy, on this account, and should be reserved for cases where the individual literally cannot make or participate in the decision in question (she or he is an infant, is unconscious, cannot be contacted, etc.). In cases where some other value is taken to be so importantly at stake as to justify overriding the individual's exercise of autonomy, appeal to hypothetical consent obscures the real issue. Instead, direct appeal should be made to the competing value and to the claim that its importance justifies circumventing consent in the case at hand.[15]

For a person to exercise autonomy effectively, the results of his or her critical examination and rational reflection must be implemented, where appropriate, in his or her actions. The implementation of decisions is an essential part of the exercise of autonomy. Failure of implementation may result from internal conditions such as weakness of will, compulsive or obsessive behavior, etc., or from external conditions such as physical or societal constraints prohibiting the appropriate action(s). Of course, such internal and external constraints interact in many ways. An obvious example is that strong external sanctions constitute potential strains on one's internal strength of will.

Indeed, an individual may not exercise his or her internal capacities in a given situation because he or she perceives the external sanctions for any but the prescribed course of action as sufficiently strong to preclude implementation of any other choice, rendering deliberation otiose.

A related source of failure to exercise one's capacities is failure to recognize the full range of options that are, or could be, available. External conditions may (accidentally or by design) render invisible—rather than merely ineligible—some or all of the potential alternatives. Or there may be a failure of imagination on the part of the potential chooser. Often there is a mutually reinforcing combination of the two.[16]

Thus, one's autonomy, one's control of one's life, is suppressed or undermined if one is prevented or discouraged from exercising it in situations where the choices made (or not made) and implemented (or not implemented) significantly affect the direction, shape, quality, or substance of the life one lives. It is worth noting that these include vast numbers of small, undramatic, everyday kinds of situations as well as the relatively few more obviously momentous potential turning points or crossroads in our lives.

To have a decision that importantly affects one's life made and implemented without one's participation tends to undermine one's autonomy even if the decision made is the choice one favors or the choice one would favor if one went through the process of deliberation. Suppose you unreflectively favor option A, but, if given the actual choice, you would reflect and, having reflected, would choose B. Now suppose I impose A on you, pointing out that it corresponds to your own preference. Perhaps you will be satisfied, but your autonomy will have been undermined in two ways. You will have lost the chance to exercise your capacities for deliberation and choice in this matter, to exercise your autonomy. And your life will be shaped differently from the way you would have shaped it had you acted autonomously. You will live out option A, whereas, had you actually chosen, you would have picked option B.

Suppose, instead, that I impose option B on you, insisting that, if you reflected critically and rationally on the matter, you would choose B, even though you now favor A. You may, with some justice, respond that the only way to be sure of that is for you to go through the process and choose for yourself. You might add that the value of having the option one chooses is not exhausted by getting the same option some other way. That one has chosen it is part—and in some cases the largest part—of its value. You will object, rightly, that your autonomy has been undermined in this situation.

Now suppose that, persuaded by your arguments, I invite you to deliberate for yourself. You do, and, much to my chagrin, you do not choose option B after all. I now say I'm sorry, but I must impose option B anyway. Perhaps I still insist that you would have chosen B if only you had deliberated more carefully, had more information or a better understanding of it, etc. Or perhaps I grant that you have chosen A (or C, an alternative I had not envisioned) despite my belief that B is the better option. I listen to your reasons for your choice and am not persuaded. So I overrule your choice and implement B. Suppose, further, that I now claim that your autonomy has not been violated or undermined, since you have participated in the decision-making process. We have discussed the matter; you have exercised your capacities for deliberation and choice and have provided input that I have listened to and taken into account in making the final decision.

You object more loudly than before. You claim that I invited your participation in bad faith—that I never had any intention of implementing any plan but B. You claim that your participation was strictly pro forma, all form and no substance, since I had, and retained, the power to implement the option of my choice regardless of your input. You feel that I have used you and your good faith efforts in order to give my unilateral choice the appearance of democratic legitimacy. Even if I sincerely claim that I was prepared to change my mind if you could persuade me that some other option really was better on the whole, you will be justified in claiming that you were prevented from exercising your autonomy due to the inegalitarian nature of the situation.

Yet autonomy does not require that one always get one's way. Recall that, on the present account, the process of deliberation and decision is not, in the first instance, an internal process that can, by extension, be externalized so as to involve two or more persons. Just the reverse is true. It should be clear, then, that one does not necessarily surrender one's autonomy, on this conception, by submitting to a group decision one does not favor, so long as one has participated as an equal in the process of deliberation and choice.

This conception of autonomy, then, does permit some appeals to the notion of indirect consent. In cases where the individual voluntarily joins a group and agrees to abide by group decisions on certain matters, where the individual actively participates in formulating and evaluating the alternatives and in making and implementing the decision, indirect consent is a robust form of consent that is fully consistent with the value of autonomy. (One may, of course, withdraw indirect consent in some cases, just as one may sometimes withdraw direct consent.) Where these conditions are less than fully satisfied, however, appeals to indirect consent become tenuous, increasingly so as the degrees of active, voluntary, egalitarian participation in the decision-making process diminish.

Although, as we have seen, autonomy does not require that one always get one's way, there are some decisions over which autonomy does seem to require that each individual have effective veto power. Suppose that B would like to become lovers with A but A does not reciprocate. Although A's negative decision may importantly affect B's life, it is A's decision to make. B would, of course, have the same veto power if the situation were reversed. Or suppose that C and D are lovers. C plans to go on a rock-climbing expedition and very much wants D to go with her. Despite her commitment to C, D may decline to participate in such a risky activity. Yet, if she does consent to go, she may have to make commitments to C—and to others—that will obligate her to take risks she would not otherwise take. Agreeing to participate in the expedition may entail agreeing to abide by group decisions on risky matters and to undertake dangerous rescue efforts in the event of an accident, for example.

What distinguishes those cases where individual veto power is appropriate from those where it is not? Unfortunately, the answer to this question is far from clear. There seem to be several factors involved, and one would need an account of the role and the weight of each. To complicate matters further, these factors may overlap and interact in complex ways. Moreover, even persons who agree on the general conception of autonomy being developed here may disagree about what some of these factors are, their relative importance, or their implications in particular instances. Perhaps the best thing to do at this point is simply to mention some of these factors and then turn to our three policy arenas and see what we can learn about these factors by considering their role in our responses to practical issues.

One factor is the nature of the bond(s) or association(s) among the persons likely to be importantly affected by the decision in question. To what extent, if any, is it a voluntary association? To the extent that it is not voluntary, what is it that makes them a group? What sorts of relations, understandings, and commitments does the association involve? Within the group of those likely to be importantly affected, of course, there may be numerous subgroups, and the nature of the relationships within and across such subgroups may vary.

Two other crucial factors seem to be what is at stake and for whom. Do the stakes involve potential threats to such fundamental concerns as life, health, and bodily integrity? Do they involve threats to important values such as community, freedom, or autonomy itself? Are the stakes the same for all those likely to be importantly affected? Are they comparable? What is the relationship between how the decision gets made and who stands to gain or lose what?

Air Pollution: The Three Arenas

Secondhand Cigarette Smoke

In this arena, several considerations seem to me to support a strong presumption—on both personal and policy levels—in favor of individual veto power as against allowing group decisions to override an individual's choice. For one thing, in this arena (as contrasted, for example, with long-range transported pollutants) such an approach is quite feasible. Indeed, in many cases it could be implemented without interfering with the pleasure and convenience of those who choose to smoke.

Involuntary exposure can be virtually eliminated in many public places by a reasonable combination of measures such as adequate ventilation plus sensibly located smoking and nonsmoking sections in restaurants, train and bus stations, large airplanes, employee lounges, workstations, and so on. To a large extent, the wishes of both those who choose not to be exposed to tobacco smoke and those who choose to smoke can be accommodated at relatively little cost or inconvenience. Where it is possible, this seems, from the point of view of autonomy, the best course and the course that public policy ought to take.[17]

But what about places where involuntary exposure cannot reasonably be prevented by such measures: elevators, buses, small waiting rooms, classrooms, meeting rooms, and so on? And what about places where public policy on such matters does not, and we feel should not, reach: private homes, automobiles? In the former case, I would argue that, as a general rule, smoking should be prohibited, as indeed it already widely is in, for example, buses and elevators. In public places where ventilation and separate sections are not feasible, this seems the only workable way to prevent involuntary exposure. For reasons I'll come to in a moment, public policy ought to protect persons from involuntary exposure to tobacco smoke. In the latter case, it seems to me that, as a general rule, persons (morally) ought to give individuals veto power. That is, persons ought not to expose others to secondhand tobacco smoke without their explicit (and informed) consent. A number of considerations seem to me to support these conclusions, as well as to help account for possible exceptions.

Let us look at what is at stake and for whom in a decision on whether cigarette smoking shall occur in a particular enclosed space. First, of course, what is at stake for a person who would be involuntarily exposed in the event of a positive decision is of a different nature from what is at stake for a smoker in the event of a negative one. As we learn more about the short- and long-term effects and hazards of exposure to secondhand tobacco smoke, it becomes clear that what is at stake for the nonsmoker is not merely aesthetic displeasure at an offensive odor or moral offense at observing someone indulging in a "nasty habit." Instead, what is at stake is involuntary exposure to a combination of air pollutants that can cause short-term effects ranging from headache and watery eyes to acute angina attacks, and that pose long-term risks of disease, and even shortened life span.[18] A commitment to autonomy—to people controlling their own lives—would seem to require that persons not have such discomforts and risks unnecessarily imposed on them without their consent.

At stake for the smoker, on the other hand, is loss of the pleasure and convenience of smoking when and where he or she pleases. Additional discomfort is involved for the habitual smoker who must refrain for longer than usual. It can be argued, however, that this discomfort is self-imposed to whatever extent the addiction that causes it is self-imposed. While this does not lessen the discomfort, it does differentiate it from that of the nonsmoker, which is entirely imposed by others. Further, no short- or long-term health risks are associated with temporarily refraining from smoking—just the contrary. So, while a commitment to autonomy may require that persons who choose to smoke not be prohibited from doing so (but see Note 17 for a qualification), it does not require that they be permitted to expose others to the discomforts and risks of cigarette smoke without their consent. Nor does it require that they be protected from the discomfort involved in refraining from doing so.

It is worth noting, as well, that the stakes are not the same for all nonsmokers. Individual physical reactions to tobacco smoke vary widely. Those who would suffer most acutely are likely to be in an extreme minority in most groups. Thus in a system of pure majority rule, with no individual veto power, such sensitive individuals would have no assurance of protection. In a society in which a significant percentage of the population smokes, without policies to prevent involuntary exposure, such an individual's life would be excessively (and unnecessarily) uncomfortable.

In addition, of course, prolonged or intense exposure to secondhand smoke can be very uncomfortable and debilitating even for persons who would not be characterized as extremely sensitive. Hence, the more heavily outnumbered the nonsmokers in a group, the more intense will be their exposure if the group is permitted to override their veto.

These considerations support, as a general rule, giving individuals effective veto power in this matter. At the same time, they enable us to account for the view that a nonsmoker who suffers no short-term effects from tobacco smoke and who is utterly unconcerned about the long-term risks of secondhand exposure would be somehow abusing such veto power if he or she invoked it motivated solely by aesthetic or moral displeasure, by malice against a particular smoker or against smokers in general, etc. A commitment to the autonomy of the nonsmoker does not require that he or she be protected from these sorts of offenses to his or her sensibilities, nor that he or she have the power unilaterally to prevent others, even temporarily, from harming themselves.

Let us briefly turn our attention now to the nature of the association among the persons likely to be affected by the decision in question. In particular, to what degree is the association voluntary? It might be tempting to say that, in the vast majority of instances, the association is entirely voluntary and that this fact vitiates the above arguments for veto power. The individual, it may be claimed, is at liberty to stay away from or to leave places where people smoke. But this will not do. He or she must work, travel, eat, attend classes and meetings, and so on, like everyone else. Many individual instances, e.g., of eating in a particular restaurant or taking a particular course, may be seen as strictly voluntary when viewed in isolation. But in a society in which a substantial proportion of the population smokes, avoiding involuntary exposure by eschewing all "voluntary" associations where one might encounter secondhand smoke would make it impossible to lead anything like a normal life.

Still this is a matter of degree. If a smoke-sensitive person's friends smoke, it is reasonable to expect them sometimes to sit in the nonsmoking section in order to share a meal with their friend, to refrain from smoking in a car with that person, and so on. On the other hand, if the friends do, on occasion, want to go someplace where they can enjoy their smokes, the nonsmoking friend—so long as he or she is forewarned—is indeed free to decline the invitation and appears to have no grounds for complaint on that account.

Again, however, it is important to be alert not just to isolated instances, but to patterns. It has been noticed that, if a woman's professional colleagues regularly socialize together at establishments that exclude women, she is excluded not only from the socializing, but from important professional contacts and decisions as well. Similarly, if one's colleagues regularly socialize in smoke-filled rooms, one must choose between being systematically (even if unintentionally) excluded and being exposed to secondhand smoke as a result of one's "voluntary" associations.

The more genuinely voluntary the association, then, the weaker the case for individual veto power appears to be. But when we think of voluntariness in terms of larger patterns of behavior, we see that many relevant associations are far less voluntary than they may appear when viewed in isolation.

Workplace Air Pollution

It has sometimes been argued that the only way for society to respect the autonomy of workers is for government to refrain from interfering in the relations between employers and employees.[19] Concerning workplace risks, including airborne risks, it has been claimed that workers consent to the risks they face by accepting and remaining in their jobs. Indeed, labor market behavior is taken by some to provide the clearest measure of the value workers themselves place on their lives and health.[20] The idea is that workers receive wage premiums for risky jobs, and the size of the premium accepted for a particular risk or set of risks reveals how workers value their lives, health, and safety.

But the very existence of a wage premium for risky work may be doubted.[21] One recent study finds evidence of a wage premium in otherwise relatively desirable jobs, i.e., those that are relatively highly paid, secure, unionized, etc., but not in those jobs characterized by low pay, insecure employment, lack of unionization, etc.[22] The latter, largely service-type jobs are disproportionately held by minority, women, and other disadvantaged workers. While workers in this category change jobs fairly frequently, they seldom are able to break out of these kinds of jobs into the more desirable sort. The results of this study strongly indicate that these workers are not compensated by wage premiums for occupational risks and suggest that the risks themselves may be greater in some of these jobs than in the better-paid and perhaps more obviously risky jobs. To infer from their acceptance of risky jobs that these workers consent to increased risk with no compensation is surely problematic, especially in view of the restricted alternatives available to them. Moreover, where the risks associated with a job are not obvious (as airborne hazards often are not), workers may not even be aware of the risks involved when they accept (and continue to work at) a job. Inferring consent from labor-market behavior, then, presupposes that workers are fully informed of the hazards they face before accepting (and while working at) jobs, that they have reasonable alternatives available to them if they find the risks unacceptable, and thus that they are satisfied with the wage/risk packages they currently get. Each of these assumptions is subject to serious question. In addition, the inference may rest on a failure to distinguish consent *in* a market from consent *to* a market.[23] That is, even if one accepts greater workplace risk in exchange for a higher wage, one may believe that health and safety ought not to be subject to the market in this way. In such a case, the consent evidenced by acceptance of the trade-off is very thin. It is, at best, consent only insofar as the background condition — that health and safety are treated as market commodities — is taken as a given.

Thus we should not be content to rely on the market to ensure that workers do in fact consent to the airborne hazards to which they are exposed. Nor, I claim, should we accept the view that respect for worker autonomy requires that this issue be left to the market. On the contrary, I shall argue that a commitment to autonomy can give rise to specific requirements for government action.

As Peter Railton notes in Chapter 5, the laissez-faire or noninterventionist view is typically associated with a moral theory based on a Lockean system of natural rights. This view is thought to show respect for autonomy by giving each individual full and final say over what goes on within the boundaries of his or her person and property. Railton demonstrates, however, that such a theory faces formidable difficulties in attempting to deal with the moral issues raised by the imposition of risk through air pollution. Moreover, Railton argues that such theories provide the basis, not for laissez-faire, but for very strict government regulation of polluting activities. Thus the moral foundation that natural rights theories have been thought to provide for noninterventionism in this area seems to crumble under careful scrutiny.

In addition, recall that autonomy, on the conception being developed here, is not a quality or condition that can be presupposed as inherent in persons, so that all government has to worry about is not interfering with its exercise. Rather, the development and exercise of autonomy can be helped or hindered in a variety of ways by social relations, policies, and institutions which may in turn be influenced by government in a variety of ways and which government cannot help but influence both by its actions and by its inactions. The question of what kinds of government policies best express a commitment to autonomy does not have a single and automatic answer according to this view. It is necessary to consider what the effects would be on both the internal and external conditions for autonomy—on people's ability to control their lives—of the different policy alternatives. I shall argue that government regulation is required in order to ensure certain necessary conditions for individual and collective worker autonomy with respect to airborne hazards on the job.

It is not possible to specify sufficient conditions for autonomous decision making either in general or in relation to airborne hazards in the workplace. One cannot hope to foresee all the ways in which autonomy might be undermined in an infinite variety of possible circumstances. We can, however, say what some necessary conditions are. We have discussed some necessary conditions for autonomy in general in the subsection on conditions for autonomy in the preceding section. What, then, are some necessary conditions for autonomy with respect to risks from workplace air pollution? Not surprisingly, in view of the relationship between autonomy and consent, these include necessary conditions for informed consent to such risks. In addition, however, they include conditions that depart from informed consent, as this notion is generally construed, in several ways. They involve participation in collective, as well as individual, deliberation and decision. They call for active participation in the formulation of alternatives and in the determination and implementation of both general policy and the specific decisions that take place within the framework shaped by such policy. By contrast, informed consent is generally construed in largely passive terms as may be seen, for example, in Chapter 8 in this volume. There the potential giver of consent is conceived as (a) a solitary individual concerned only to advance his or her own interests, (b) the (passive) recipient of a preestablished list of available alternatives, and (c) someone with no role in shaping the implementation of the decision once consent is granted.[24]

A commitment to worker autonomy with respect to airborne (and other) workplace hazards requires that individual workers be guaranteed the following rights:[25]

(1) The right to know
  (a) the generic names and potential hazards of the substances to which they are or may be exposed in the workplace,

(b) the specific dangers and recommended safety procedures associated with their particular jobs,
(c) the full contents of their employer-held medical records, including results of tests monitoring their body burden or other signs of exposure to any known or potential toxins that the employer tests for,
(d) statistical results of such monitoring of the work force as a whole, as well as results of ambient monitoring of the levels of toxic substances in the work environment.
(2) The right to refuse without penalty—either in terms of disciplinary procedures or lost wages—a specific assignment believed by a worker to constitute a serious threat to her or his life, health, or safety.

In addition, workers in a given workplace collectively must be guaranteed the following rights:

(3) The right of access by their collective bargaining agent to all information listed above in (1) (a)–(d) with the proviso that personal identification be removed from individual medical records. Access to such information is essential if the workers, through their union, are to bargain effectively for adequate health and safety provisions in their contract, as well as to carry out other important collective functions such as doing, commissioning, or cooperating in relevant independent epidemiological and other research, lobbying for needed governmental standards, and so on.
(4) The right, by majority vote of the workers at risk, or by decision of a worker-run health and safety committee, to shut down without penalty—either in terms of disciplinary procedures or lost wages—any operation the workers collectively deem a serious threat to life, health, or safety until the hazard is corrected or a full investigation by a qualified government inspector determines that there is no danger.
(5) The right to participate fully and meaningfully in the formulation, determination, and implementation of workplace policies concerning health and safety. This is not just a right to be consulted, but to have an effective role in making and implementing decisions.
(6) The right to be informed well in advance and to participate in decisions concerning proposed new processes or substances to be introduced into the workplace. Again, this is more than a right to be consulted.

The autonomy of workers is more directly and fundamentally undermined by the absence of these rights than is the autonomy of employers by their presence. We are talking about who determines the conditions under which workers spend some 40 or more hours per week for some 40 or more years of their lives—conditions that are bound to have profound effects on the lives of those who work in them—including, but not limited to, effects on both the quality and quantity of life resulting from health and safety conditions. If autonomy is the ability to control one's life, and if workers are to be autonomous, then surely they must control these conditions.

To those who would claim that worker autonomy is guaranteed by the right of a worker to refuse or quit any job that he or she doesn't like, for whatever reason, let me remind you that whether refusing/quitting constitutes a viable choice depends on what other alternatives are actually available. The presence of genuinely acceptable alternatives among the options open to one is a necessary external condition of autonomy. Thus the fact that a worker doesn't quit a job does not necessarily indicate that he or she finds the risks (or other features) associated with the job acceptable; it may simply indicate that no acceptable options are open to this person at this time and place with these skills.[26] Precious little is gained in terms of autonomy by bouncing around from one low-paid, unsafe, insecure, and demeaning job to another. If our society is to make good on its proclaimed commitment to respect and promote the autonomy of all its members, then it must include within its area of concern the conditions under which workers (i.e., most of us) spend half or more of their waking adult lives, and how those conditions are determined. For government to guarantee workers the rights mentioned above as well as minimum health and safety standards concerning known hazards is not paternalistic. Such provisions are necessary to ensure that workers not be forced by the absence of alternatives to accept and remain in jobs that fall below a floor of decency that must be collectively determined and enforced if it is to be enforced at all. The individual worker alone is powerless to alter the options in all but a few respects.

In closing this section, it must be noted that not one of the rights listed above is fully provided for by current federal regulations. As mentioned in Chapter 2, there are a few fairly strong local and state "right-to-know" laws, but federal OSHA is in the process of promulgating an unacceptably weak regulation which is intended to preempt state and local laws, despite provision in the OSH Act that states and municipalities may pass regulations more stringent than federal OSHA's. Even the much weaker rights that workers do have under current regulations have been bitterly fought by employers, as noted elsewhere in this volume by Punnett, Baram, and MacCarthy. Thus it is unrealistic to hope that worker autonomy will be respected or promoted in the absence of government intervention. Hence, if I am correct—or even close—a great deal more, rather than less, intervention is called for if we are genuinely committed to autonomy.

Outdoor Air Pollution

This seems the least likely of our three arenas to offer any significant role for individual consent. Even what might initially appear clear instances at least of implicit consent—such as buying a home in a polluted area, attending a conference in a polluted city, or camping where rain has the pH of household vinegar—do not hold up under scrutiny. As Peter Railton notes, "Someone who voluntarily moves into a high-crime (or high-grime) neighborhood may have acted unwisely, but he has not laid down his rights, and those who invade his person or property violate these rights."[27] Although Railton is speaking specifically in terms of natural rights theories, this point seems valid whatever account one may give of rights. If persons have rights not to be harmed in certain ways or subjected to certain sorts of risks they do not give up those rights by going into an area where such rights are protected only poorly or not at all. Persons who live, work, and travel in unsafe neighborhoods have the same rights not to be mugged as those who frequent safer environs. As Judith Thomson points out in Chapter 6 of this volume, in consenting to walk down an unsafe street, one does not thereby consent to the street's being unsafe. One does not, in other words, consent to being at heightened risk of being mugged. And if one is mugged on that street, one has not, by being there, consented to being mugged. If it is wrong to expose persons to airborne hazards without their consent, it will be no defense to say that a person entered the exposure region voluntarily (so long, of course, as the exposure region is one the person is entitled to enter).

Thomson's point holds even in cases where a safer alternative is available, and one chooses, after weighing the risks and benefits, the riskier option. *Accepting* a risk by choosing an option that entails that risk is not, it turns out, tantamount to *consenting* to that risk. Further, as Fischhoff insists, accepting that risk does not entail that one considers that risk "acceptable."[28]

Voluntary presence in a risky area can even less plausibly be taken to signify consent to the risk if the available alternatives are significantly restricted. Poverty, for example, may preclude purchasing or renting a home in any but the least wholesome neighborhoods. One no more consents to the air pollution than one does to the rats and roaches or the lack of heat and hot water. The point here is this: even if one lived in these conditions by choice—out of a commitment to live "among the people," or to devote one's financial resources to helping feed the hungry or to some other cause or project—one would still not consent to those conditions by making that choice. But it is doubly wrong to infer consent on the part of a person for whom these conditions represent the best of the available alternatives.

Two additional considerations further undermine any inference of consent. First, given the many sources and kinds of outdoor air pollutants and the long distances some of them travel, one's available options will probably include exposure to emissions from a local chemical factory, lead and carbon monoxide from heavy traffic, acid precipitation from distant coal-fired power plants, vented radioactivity from a nearby nuclear power plant, pesticide drift from spraying of nearby fields or forests, or — more likely — some combination of these and other pollutants. With such a range of available alternatives, it again seems doubly wrong to attribute consent to an individual who opts for a residence or vacation spot that will expose her or him to one such set of airborne hazards rather than the others.

Second, as in the workplace and cigarette-smoke settings, for consent to be genuine, it must be adequately informed. Outdoor air pollutants are, as Chapters 3 and 12 make clear, often extremely difficult to identify, measure, and trace, even when the dangers associated with them are well established. An individual with little or no specialized training would, as things currently stand, be unlikely to be sufficiently well informed about the potential airborne hazards of various locales so that a decision to visit or settle somewhere could plausibly be interpreted as constituting informed consent to those hazards.[29]

Outdoor air, then, is a setting in which the notion of individual consent to risks has little to offer in the way of guidance, and we must look to the underlying value, autonomy, to see what it can provide.

What the conception of autonomy developed in this chapter tells us most clearly, it seems to me, is that there can be no substitute for actual, direct, informed, and meaningful participation in the formulation, adoption, and implementation of public policy concerning air pollution by the persons whose lives, health, and values are potentially at risk. Because autonomy exists only in and through being exercised, no formula, whether of risk-benefit or any other sort of calculation, can replace the actual involvement of those potentially affected by the policies in question. What we need, if we are to respect and promote autonomy, are not more sophisticated algorithms for making decisions, but better ways of involving those whose lives will be affected by those decisions in making them. Invoking the notion of hypothetical consent to justify decisions on the grounds that those affected *would* consent to them *if* they could be consulted, were adequately informed and rational, etc., is a way of circumventing and diminishing autonomy, not of respecting and promoting it.

This is a place where theory cannot substitute for practice. It can tell us why that is, perhaps, and why the practice should be of one sort rather than another, but there is no theoretical device that can do it for us. There is no theoretical way to duplicate the actual dynamic interaction of different persons with distinct points of view, where each is open to modification in response to new information, and each has the full resources of his or her own history and experience from which to respond. Different perspectives can be brought together only by bringing different persons together, and the outcome of their interaction cannot be preprogrammed; there must be room for the creation of new, previously unimagined alternatives.

In a recent article, Roger E. and Jeanne X. Kasperson discuss the problem of how to arrive at justifiable policy decisions concerning tolerability of risks.[30] They endorse a version of a hypothetical consent model as the criterion of sound decision making.

> Sound decisions on risk tolerability require a process consistent with western democratic theory yet directive to the risk guardian. Since the public cannot hope to inform itself about and to participate in each of the innumerable decisions on risk, it delegates discretion to the risk guardians—the legislators who pass laws and the regulators who implement them.
>
> In a democracy, what these risk guardians want to know in order to make value-laden decisions is what the outcome of a considered public decision process *would be in an idealized state* where interests have become clear, the publics highly informed, issues dissected and debated, individuals free to choose, and public officials responsive to the electorate. It is a *hypothetical* state, of course, for a modern democracy cannot realize such requirements for risks or for any other public issue.[31]

At the same time, they call for measures to increase significantly actual direct participation in decision making by the potential bearers of proposed risks. The precise nature of the relationship they envision between their hypothetical model and the increased participation they call for is not entirely clear. Our earlier discussion of the limitations of hypothetical consent models suggested that the only way to determine what would be the outcome of a process under the conditions the Kaspersons describe is actually to go through the process under those conditions (or conditions approximating them as closely as possible). Perhaps the connection they see involves an intuition to this effect, together with an explicit sensitivity to the skepticism and resentment generated by the imposition of risks on people in the absence of democratic procedures. "Effective risk decisions depend," they maintain, not only upon the adequacy of the scientific information employed and the content of the decisions themselves, but "equally upon the processes and upon the institutions responsible for the judgments that emerge."[32]

In any event the concerns and conditions that they articulate as "responsibilities to risk bearers" that must be satisfied by any justifiable decision-making procedure correspond closely with those that emerge as essential if we are to respect the autonomy of potential risk bearers:

> First, there is the obvious need to inform those who will bear the risks of the nature, levels, distributions, and associated uncertainties of the risks as well as opportunities for their control. Specifically, the limits of scientific understanding should be noted. This informing function should exceed that which routinely occurs in public meetings or in Federal Register notices; it should strive for a high degree of understanding among those who will experience the risk. It is important to recognize that the risk imposer (i.e., the developer or technology sponsor) has a conflict of interest in developing and dispensing information.
>
> Second, potential risk bearers must receive ample opportunity to participate at all stages in the risk decision process. All too often, risk bearers become involved only in the final stages, after assumptions have been set, alternatives narrowed, many key decisions made, and decision makers committed to a course of action.
>
> Third, opportunities for participation are meaningless if the fundamental asymmetry of capability and resources (noted in our premises) is not addressed.[33] At the heart of risk-bearer participation is the development of an independent technical and financial capability to develop information, examine technical documents, challenge assumptions, and explore (and perhaps formulate) policy alternatives.
>
> Fourth, the decision process will necessarily allocate the burden of proof among decision participants. All too often, a particular decision or course of action is adopted and affected individuals are allowed to challenge it. Given the asymmetry of power which prevails and in order to err on the side of avoiding harm rather than enlarging benefit, the burden of proof should fall on the risk imposer.
>
> Finally, the risk bearer must have access to some means of redress by which a decision may be appealed to a higher body. Although this adds a further potential delay in decision making, it is an essential ingredient of due process for those who will experience the harms.[34]

While I am less sanguine than the authors appear to be about the prospects for fulfilling these responsibilities within the framework of a society so structured that it must rest content with a hypothetical model of democratic decision making, I am perhaps more optimistic in another respect. I am not convinced that "a modern democracy cannot realize such requirements [as those outlined in their hypothetical ideal] for risk or any other public issues." Further, if democracy is about people governing themselves, about people controlling their own lives, about autonomy, then such a view would seem to entail that democracy and modernity are incompatible. While there is much in contemporary life to support this claim, I remain hopeful that more genuinely democratic institutions and practices are possible. If we are committed to respecting and promoting the autonomy of all members of our society, then we must commit ourselves to developing and implementing such institutions and practices, despite the evident difficulties involved. No other approach promises better, less controversial, or less divisive decisions. Thus, I believe I have shown that this approach is to be preferred on grounds of the intrinsic value of autonomy (Scheffler's second conception of the moral importance of consent). I believe, but have not argued, that it is preferable also on grounds of the general good (Scheffler's first conception). That is, I believe that this approach will lead both to enhanced autonomy and to better decisions and a better society in the long run.

Social-historical patterns of human action and inaction result in institutions, practices, policies, and laws that shape the kinds of alternatives that are and can be made available to people. These social conditions largely determine the extent to which people—individually and collectively—can control their own lives. Thus they affect both the degrees of autonomy generally attainable by the persons and groups choosing among the alternatives and the degrees of autonomy that can plausibly be said to be expressed by their specific choices. If we care about autonomy, we must be prepared to examine our institutions and practices and to alter those that tend to undermine or interfere with it.

Notes

1. Samuel Scheffler, "The Role of Consent in Legitimating Risky Activities," this volume, page 76.
2. R. S. Peters, "Freedom and the Development of the Free Man," in *Educational Judgments*, edited by James F. Doyle (London: Routledge and Kegan Paul, 1973), p. 123.
3. Another thing I cannot hope and shall not even try to do is to sort out the conceptual relations among autonomy and its close relatives, such as freedom and authenticity, or its alleged enemies, such as manipulation and coercion. Just to canvass the recent philosophical literature on these topics would be an enormous undertaking.
4. Immanuel Kant, *Foundations of the Metaphysics of Morals*.
5. "Sex" designates the biological differences: male/female; "gender" designates the social differences: man/woman.
6. Jean Piaget, *The Moral Judgment of the Child* (New York: The Free Press, 1965). What follows is not a systematic exegesis, but an impressionistic account of some aspects of Piaget's work that I have found suggestive.
7. In stating that autonomy is a fundamentally social phenomenon, I not mean to be claiming either that autonomy always is or always ought to be exercised in a social, rather than individual, fashion. For present purposes, I shall be content to try to support the following claims: (1) Autonomy is fundamentally social in its origin and development. (2) The development and the exercise of autonomy are inextricably interrelated. (3) Autonomy can be exercised in a social fashion.
8. Piaget, *The Moral Judgment*, p. 251.
9. Ibid., p. 95.
10. Ibid., p. 403.
11. I cannot here attempt to give an account of rationality. Elsewhere, I argue against the "value-neutral" conception, currently standard in economic theory and widely adopted by

philosophers, and begin to develop an explicitly normative conception that I hope eventually to work out more fully in conjunction with the notion of autonomy being developed here. (See Mary Gibson, "Rationality," *Philosophy & Public Affairs* 6 [Spring 1977]: 193–225.)

12. Even if one occasionally "breaks down" and lets value *B* win out, the judgment that value *A* is weightier and should win may stand. If one always, or very often, lets *B* win, then the validity or sincerity of one's judgment must come into question.

13. Does this openness to question, criticism, and revision preclude genuine commitment of the sort necessary for there to be a self in the sense required for autonomy? I think not. That nothing is, in principle, immune to question does not mean that everything is always "up for grabs." As Joel Feinberg says,

> Rational reflection . . . presupposes some relatively settled convictions to reason from and with. If we take autonomy to require that all principles are to be examined afresh in the light of reason on each occasion for decision, then nothing resembling rational reflection can ever get started. [Feinberg, "Idea of a Free Man," p. 26]

If there is a problem here, it seems to me broader than, but not otherwise different from, familiar epistemological problems. The impossibility of refuting radical skepticism guarantees that no item of knowledge is absolutely certain, beyond question. At the same time, there is no escaping the need always to hold some things constant while questioning or examining others. And some of our beliefs are far more central to our way of thinking, and hence far less likely to be revised or abandoned than others. Among those central ones are some we puzzle over philosophically without even being able to imagine actually carrying on without them (our commitment to rational thinking, our belief that there is and we are part of a physical world, and our conviction that there is a difference between coincidental and causal relations, to mention a few). These problems and puzzles do not prompt many of us to shrug and abandon our efforts to think and act rationally.

Autonomy involves critical examination of and rational reflection on, not only our factual and epistemological beliefs and values, but all of them: moral, political, aesthetic, and so on. Such examination and reflection require some relatively settled convictions in order to get started. Autonomy cannot require that we distance ourselves from all of them at once—there would be no self remaining—but it can and does require that none is in principle beyond examination and criticism. And, recognizing that achievement of autonomy is both an ongoing process and a matter of degree, we may say that actually subjecting one's most central convictions to critical examination and rational reflection is a significant element or feature of autonomy.

14. One way in which some people seem to cope with unresolved conflicts among their values and purposes is, in effect, to assign the members of conflicting sets to different selves that may overlap with respect to other constitutive features but not the incompatible ones. This seems to me an unsatisfactory solution. That the boundaries of the self are flexible does not mean that one can avoid or resolve contradictions by "wearing different hats."

15. Even where the value at stake is the autonomy of the individual whose consent (or lack of it) is at issue—for example, where paternalistic interference is said to be justified on the grounds that it is needed to protect or promote the individual's future autonomy—appeal to hypothetical consent is not to the point. What is to the point is a direct appeal to the value of autonomy and the argument that the threat to the individual's future autonomy is sufficiently grave to warrant interference with her or his present exercise of autonomy. (The present account of autonomy, with its emphasis on the idea that autonomy is an exercise-concept, should lead us to view with considerable skepticism, though not to dismiss out of hand, such arguments for autonomy-enhancing paternalism.)

16. See Chapter 8 of this volume for further discussion and illustration of this point.

17. I cannot address here the many interesting and important issues that might be raised concerning the autonomy of the smoker. But it should be said, in qualification of the statement above, that a policy that accommodates the wishes of smokers to smoke counts as desirable, on that account, from the point of view of autonomy, only if, or to the extent that, deciding to smoke is an exercise of autonomy. For a decision to smoke to be autonomous, the potential smoker must be fully informed about the risks to self and others of exposure to tobacco smoke. Thus a society committed to autonomy must be committed to ensuring that its members are fully informed about these risks.

18. These discomforts and risks are detailed and documented in Chapter 1 of this volume.
19. Adina Schwartz presents a brief and very accessible account of the history of this argument's influence on U.S. labor law in Part I of her essay, "Autonomy in the Workplace," in *Just Business: New Introductory Essays in Business Ethics,* edited by Tom Regan (New York: Random House, 1984), pp. 129–63. In Part II of that essay she offers her own account of autonomy and argues that government regulation of many aspects of employer-employee relations can be defended on grounds of worker autonomy. Despite some apparent differences—perhaps only in emphasis—there is a great deal of common ground between Schwartz's views on this subject and my own. See also Mary Gibson, *Workers' Rights* (Totowa, N.J.: Rowman and Allanheld, 1983).
20. E.g., Richard Thaler and Stanley Rosen, "The Value of Saving a Life: Evidence from the Labor Market," in *Household Production and Consumption,* edited by Nestor Terleckyj (New York: National Bureau of Economic Research, 1975). Additional references to this literature can be found in Brown, "Equalizing Differences," and Graham et al., "Risk Compensation," cited in Notes 21 and 22.
21. See Charles Brown, "Equalizing Differences in the Labor Market," *The Quarterly Journal of Economics* (February 1980): 113–34. This paper reviews and analyzes the evidence for wage premiums in the recent literature and finds it inconclusive.
22. Julie Graham, Don M. Shakow, and Christopher Cyr, "Risk Compensation—In Theory and Practice," *Environment* 25, no. 1 (January/February, 1983): 14–20, 39–40.
23. I owe this way of putting it to Mark Sagoff.
24. Fischhoff explicitly recognizes these restrictions and notes that there are areas of possible divergence between informed consent so construed and autonomy, as well as between both of these and optimal decision making. He suggests that where they diverge, policies defining requirements for informed consent may lean one way or the other depending on whether optimal decision making or autonomy is of greatest concern in the case in question.
25. As noted earlier, my use of the language of rights here should not be taken to indicate that I subscribe to a natural rights theory. My views on the nature and limits of rights are sketched in *Workers' Rights,* Chapter 5. In addition, detailed discussion of the need for and current status of the workers' health and safety rights mentioned here may be found in Chapters 1 to 3 of that book.
26. This is what Baram refers to in Chapter 9 of this volume as "structural coercion." Unlike Baram, though, I believe that this condition can and must be addressed at the policy level—that is the only level on which it can be addressed.
27. Railton, Chapter 5 of this volume, p. 93.
28. Fischhoff, Chapter 8, this volume. These two points, though distinct, are closely related and overlap somewhat with each other as well as with Railton's point just above and with the distinction noted in the previous section between consenting *in* a market and consenting *to* a market. As Thomson and Railton suggest in Chapters 6 and 5 of this volume, these points all warrant consideration also in connection with workplace risks.
29. This should not be taken to mean that efforts to educate and inform the public about the nature of the risks and the limits of scientific knowledge are not needed. People are entitled to know about the risks to which they are exposed independently of whether their presence in an exposure area is to be construed as consent.
30. "Determining the Acceptability of Risk: Ethical and Policy Issues," in J. T. Rogers and D. V. Bates, eds., *Risk: A Symposium on the Assessment and Perception of Risk to Human Health in Canada, October 18 and 19, 1982 . . . Proceedings* (Ottowa: The Royal Society of Canada, 1983). Reprinted as CENTED Reprint No. 41 (Worcester, Mass: Center for Technology, Environment, and Development, Clark University). They suggest that the term "tolerable risk" may be less misleading than the more commonly used "acceptable risk."
31. Ibid., p. 149, emphasis added.
32. Ibid.
33. The relevant premise states, "Power relations in risk decision processes tend to be asymmetric: risk imposers nearly always have superior knowledge and resources to promote the expansion of potentially hazardous technologies; risk bearers nearly always have few resources and limited and tardy access in resisting such technologies."
34. Kasperson and Kasperson, "Determining the Acceptability," pp. 149–50.

8

Cognitive and Institutional Barriers to "Informed Consent"

Baruch Fischhoff

One sign of our times is that individuals are increasingly requested to acknowledge explicitly that they are undertaking an act that entails some risk among its consequences. Participants in medical experiments sign consent forms. Temporary workers in some nuclear power plants are briefed about radiation risks before being asked to start work. Buyers of homes built before 1950 may have to agree to the risk of lead-based paint within. Diners may have to give permission to being seated in the smoking section of restaurants. In each of these cases, the individuals involved must make a choice among several options, at least one of which both involves risk and is of particular interest to another party. If they adopt that option, then the other party may wish to claim that informed consent has been given. For example, a physician may claim that signing the form constitutes granting informed consent to the risks of a new surgical procedure; or an employer may argue that accepting a job after receiving the briefing means consenting to the incumbent risks.

This chapter applies the perspective of behavioral decision theory to the question of when these claims are justified. This perspective provides: an analytical framework for characterizing human actions, a body of empirical evidence regarding when people have difficulty choosing suitable actions, and an array of practical aids for helping people make decisions in their own best interests. These three aspects of the "theory" are applied in turn. They are followed by a discussion of the limits of this approach, considering the inadequacies of behavioral decision theory, of optimal decision making as a criterion of informed consent, and of the concept of informed consent itself.

What Is Informed Consent?

In making a decision, an individual makes a choice among options. The set of options may be simple (agree to the operation, reject the operation) or complicated (agree to the operation under condition $C_1, C_2, \ldots, C_n$, reject operation in favor of procedure $D_1, \ldots, D_n$). However, it must have at least two members. If one has no choice regarding what to do, then one cannot make a decision, but must accept the single available option. But this kind of acceptance hardly constitutes consent. Decision making must also be deliberative, that is, two or more options must not only be available in principle, but an explicit choice must be made among them. If one just goes along with a single option (e.g., out of habit) without considering alternatives, one cannot be said to have consented to it.[1] In both these respects, making a decision is a necessary condition for giving informed consent.

Within behavioral decision theory, the ultimate criterion for good decision making is optimality. A decision is considered to be optimal if an individual chooses the course of option that is in his or her own best interests, exploiting fully the available evidence. "Optimality" is judged in terms of how well an individual makes use of the information available (or potentially available) at the time the decision is made, not in terms of what eventually happens. Whenever some uncertainty surrounds a decision, there is some chance that the best of all possible choices will not lead to the best of all possible outcomes. Thus, a decision to take an experimental drug cannot be faulted solely because it happened to lead to an unlikely side effect. "Best interests" is defined in terms of whatever consequences the decision maker considers to be important. Those consequences need not be restricted to selfish or egotistical ones, although they often are. Thus, one's interests may include making the doctor happy or helping to develop new drugs, and one may be ready to accept some personal risk in order to achieve those goals. "Exploiting fully the available evidence" need not mean collecting all the evidence that is available. The costs of acquiring and digesting additional information need to be weighed against the potential contribution of that information to one's welfare. Thus, a person cannot be faulted for making a snap decision if the best alternative is clear and further evidence seems unlikely to change one's mind.

Decision making is, in this view, a process involving the assemblage and mulling of information with the aim of doing the best one can. The final stage in that process is the act of choice informed somehow by the process that went before it and by the considerations laid out before the decision maker in order to serve that last stage.

A minor industry has grown up for devising different prescriptive decision rules, each of which might be used to identify the best option. One thing these rules have in common is that they look at all pertinent consequences of the options considered. For present purposes, these consequences might be categorized as risks, nonrisk costs, and benefits,[2] or as risks and net benefits (comprising the sum of all consequences other than risk). The optimal choice is determined by risk levels alone only if all other costs and benefits are either negligible or equal for all alternatives. That is, risky decisions are not about risk alone. That some level of risk is associated with the option chosen does not mean that that level of risk is acceptable. Options, not risks, are chosen. Although in some cases the risks may be of particular interest (e.g., for a regulator charged with defining "acceptable risk"), they should have no special status for the individual making the choice. Decision makers should be interested in getting the best deal for themselves, all things considered.

For the party offering a risky option, the important question is whether informed consent has been granted to it. For the potential consenter, the important question is whether that option is the most attractive option, all things considered. That individual's best interests have been served when he or she has made the optimal choice. Whether that choice is interpreted as informed consent is, to a first approximation, someone else's affair. That one option with a particular risk has been accepted is no reason to assume that another option with the same risk would or should be accepted (unless, of course, the accepting individual viewed it as having even more net benefits). Nor does acceptance confer the mantle of "acceptability" on an option (much less on its risks). That the option selected is the best available does not mean that it is liked or that it will not be abandoned should something better come along. Thus, here as elsewhere, the term "acceptable risk" is not a useful one.

From this perspective, the informed consent problem becomes one of optimal decision making. Specifically, I will advance and then explore the following thesis: if the focal option is the best one for the decision maker, then the other party may claim that its selection constitutes informed consent. The other party's task is therefore threefold: making the focal option more attractive than its competitors, guaranteeing that its virtues are recognized, and demonstrating that the decision maker has used an optimizing procedure in selecting it. The next section considers some of the cognitive barriers that limit people's ability to optimize on their own behalf. It is followed by a discussion of methods for improving decisions that might be adopted either by the individual or by the other party (in order to ensure the optimality of the individual's choice). The final section addresses ways in which informed consent problems may be more than just decision-making problems and how their consideration highlights limitations of behavioral decision theory.

Can People Give Informed Consent?
(Can they make optimal decisions?)

Prescriptive models of decision making[3] typically contain several interrelated steps, one characterization of which is:

1. List all possible courses of action.
2. Evaluate the attractiveness of the consequences that may arise from undertaking each course of action.
3. Assess the probability that each consequence will, in fact, be incurred with each action.
4. Combine these considerations so as to identify the best possible course of action.

These tasks are interrelated in the sense that some knowledge of each is required to perform the others. For example, the choice of which options to evaluate requires some idea of what consequences are worth achieving (or avoiding). As a result, a successful decision-making process is often iterative, with each pass through the stages capitalizing on the insights generated by its predecessors.

Problems with any of these subtasks increase the chances of a suboptimal choice. Yet few people receive explicit training in decision making. As the following brief review of the research literature may suggest, the frustration that people often express with making decisions may be explained by both the inherent difficulty of making decisions and their own lack of training.[4] How calamitous these problems are for people's choices will depend upon the particulars of individual decision problems. In some cases, even a clumsy, cursory consideration will lead to the same choice as would a careful, skilled analysis. In others, the best option is hard to find and perhaps obscured by the problem's surface representation, a difficulty that may be deliberately exacerbated by whoever is presenting the problem.[5] Of course, having the requisite cognitive skills is no guarantee of using them. People may not invest the effort required to choose the optimal option simply because the stakes do not justify it or because they underestimate the difficulty of making good decisions (an error that could itself be construed as reflecting the lack of a necessary cognitive skill).

Listing the Options

A first obstacle to making a complete list of options is failure to realize that there is a decision to be made. The very introduction of formalized procedures for soliciting informed consent reflects an acknowledgment that people often do not realize that they have any choice but to accept the risky option offered to them.[6] Although unreflective compliance may be exploited and encouraged by interested others, it may occur with no outside help. Two contributory factors are the cost of making decisions, which discourages reflection, and the force of habit, which encourages doing things the way they have always been done.

Even when the possibility of making a choice is recognized, convenience and habit may constrain the list of options considered. Generating good options often requires a creative act. In informed consent situations, it requires an act of imagination to realize that one's choices may not be constrained, say, to agreeing to an operation or doing without the best care available. One may be able to go to another physician, wait until the procedure has been tried on other people (whose situations are more desperate), get some additional precautions incorporated in the procedure, or obtain some side payment for compliance (e.g., a private room, subsequent briefings on the progress of the research). Such option creation requires both a generalized skill, which is undeveloped in many people, and a knowledge of the possibilities inherent in any particular situation. Knowledge of the range of available options is distributed unequally in most informed consent situations, to the disadvantage of the potential consenter. Unless this imbalance can be redressed, optimality (hence, informed consent) may be impossible.

These difficulties, as others, would be ameliorated somewhat if people had an accurate perception of their own limitations. Knowing that the set of options is incomplete, or even systematically biased, could prompt one to try harder to be creative or look for help. Unfortunately, whether due to lack of skill or lack of knowledge, one's own failures of imagination are inherently hard to detect. It seems as though people are unaccustomed to asking about the limits to option sets or identifying omissions if they try to press on those limits. Options that are out of sight also tend to be out of mind.

Evaluating the Consequences

Assessing the desirability of outcomes would seem to be unproblematic, the last redoubt of unaided intuition. Who knows better than an individual what he or she likes and dislikes? Surely, people know what they want. Although often reasonable, this assumption may be doubtful in just the sorts of situations that arise in informed consent decisions. Some preferences (e.g., for sweet things) are presumably innate. Most, though, must be learned, either through trial and error, which reveals what one does like, or through socialization, which teaches what one should like (inducing tastes which may become indistinguishable from one's innate preferences). The more novel the situation, the less likely one is to have articulated values regarding it.

In the absence of a prepared response, value questions require an exercise in inference. One must determine which of one's trustworthy basic values are relevant to a particular situation, how they are to be applied, and what weight is to be given to each. Unless one is skilled in making such inferences, it is natural to turn to the person or context posing the question for hints regarding how to respond. As survey researchers have discovered to their chagrin and documented through careful experimentation, it can be very difficult to avoid giving some clues to a respondent who is seeking them. The seeking of clues regarding what constitutes an appropriate response is a fundamental aspect of human conversation, even in situations where only one participant's feelings should be considered, and even when "conversation" is conducted via an interviewer or questionnaire that strives at impassivity.

The very act of posing a question in an interview situation may be construed as conveying a number of significant messages, each of which may subtly affect what the respondent says, even feels. These messages include: (a) that the respondent has thought enough about the topic to have articulated beliefs; (b) that the respondent should have such beliefs; (c) that the interviewer has a right to know those beliefs; (d) that the respondent's personal views are preeminent (as opposed to those views that might be evoked by a group setting); (e) that there is no room for group consultation or collective action (such as might enable the respondent to get a better deal from the interviewer); (f) that the interviewer (also) appears as an individual and not as the representative of some external organization.

Further cues may be found in the sex, race, or social status of the interviewer (one has different self-perceptions in front of different people, and different stories to tell), in the number of questions posed on the topic (allowing different degrees of depth and rumination in one's thinking), in the specific words used (the associations evoked), in the interviewer's haste or leisure, in the order in which questions are asked, in the questions that are not asked, in the degree of implicit pressure for maintaining consistency among different responses in the interview or between interview responses and previous behavior, or in the interviewer's apparent ability to ascertain and interest in ascertaining the respondent's candor. Such influences pose a threat to the validity and usefulness of any interview that attempts to divine "what the respondent really thinks." Clearly, each offers some manipulative possibility for an interviewer who wishes to elicit a particular response, whether a pollster trying to show the popularity of a candidate or a physician trying to show that patients do not mind a particular treatment.

Three brief examples may give the flavor of some of these elicitation effects: (a) People are less likely to pay to protect themselves against a possible loss if the payment is labeled a "sure loss" rather than an "insurance premium." (b) A medical treatment program will seem less attractive if described in terms of the percentage that will live. (c) More people report being happy with "all things taken together" on surveys in which the preceding questions deal with married life than when they deal with other topics.[7]

The survey researchers' response to these problems is twofold. First of all, there is an attempt at standardization.[8] The interviewing context is kept as similar as possible for all respondents, thereby keeping the unintended influences as constant as possible. If successful, this strategy reduces the random error in questioning, while accepting some (typically unknown) constant error. It allows one to compare the answers of different respondents, without, however, being certain of the validity of any one. As a result, standardization is a necessary but not a sufficient condition for informed consent situations.[9]

The second, and complementary, strategy is to try to weed out unintended influence as much as possible. This has been an important goal since early pollsters discovered obviously biased procedures such as suggesting that everyone (of any virtue) responds a particular way, using emotive language in describing the response options, or failing to offer response options allowing certain views to be expressed.[10] As a matter of intellectual hygiene and professionalism, reputable survey researchers diligently avoid embedding such pressures in the questions they ask. This approach faces three difficulties, however; one is practical, the other two are fundamental. The practical difficulty is that the research based on the prevalence and magnitude of context effects is fairly limited, and detailed study is needed to ascertain what effect particular variations in questions will have. For example, there is good empirical reason to believe that the race of the interviewer affects response to race-related questions (views are more extreme and more prejudiced with same-race interviewers), but not to other questions (e.g., choice of favorite entertainer).

Even if an adequate (even complete) evidentiary base were available, however, it would not, in itself, indicate the correct way to pose questions. In order to make use of the research, the interviewers (or those who employ the interviewers) must know exactly what question they want to ask. Do they want to know what whites tell blacks about racial questions or what they tell other whites? Do they want reflective or spontaneous evaluations of a particular issue? Is the aim to predict behavior in white-dominated workplaces or in integrated political groups, to predict purchases of toothpaste or decisions to marry? Answering these questions is essential to the design of an interview. It requires an analysis of the purpose to which the answers will be put. A hard look at the survey results reported in the popular press suggests that such clarity of purpose is not always present; questions are often asked just because it would be "interesting" to know the answers. Clarity of purpose is perhaps most common when questioners have a vested interest in eliciting a particular response and hope that neither respondents nor readers of those responses will detect the biased framing.[11]

The final impediment to even a wholehearted attempt to ask the right question is the possibility with which this section began, namely, that the respondent does not have an articulated answer. In practical situations such as those involving informed consent, the questioner often cannot take "no answer" for an answer. But having a question, even the right question, carries no assurance that the respondent will have a trustworthy answer, particularly given the limited time for reflection that most interviews afford. Indeed, the care that is required to formulate the right question might be taken as an indicator of the difficulty that a respondent will have in understanding it in all its subtlety (i.e., why it is phrased the way it is and not in other superficially similar ways). A related form of deep ignorance is not understanding in some essential sense what an unfamiliar consequence of one of the options is like. Can a healthy person know what it means to have the pain of cancer or the anxiety of angina? Might they overestimate the inconvenience of other maladies? Juvenile debates over whether it is worse to be blind or deaf reflect such strongly held but experientally uninformed opinions.

The life-and-death questions that lie at the core of many informed consent questions would seem to be particularly likely to involve inchoate and manipulatable values. Society has just begun to think about these issues in a reasoned fashion. How well can individuals be expected to know just what they want in novel-yet-consequential situations?

Assessing the Likelihood of Conseqences

Not all decisions, or even all informed-consentlike decisions, involve uncertainty. Examples include deciding whether to work in an asbestos mill where, by some accounts, everyone's life is shortened, with the only uncertainty being about how long.[12] Uncertainty arises when it is possible to say "this *may* hurt (or please) you." Such statements are, of course, singularly uninformative unless the likelihood of pain (or pleasure) can be given more precisely. Once they are, decision makers will want to discount the consequences accordingly.

The problem created by uncertainty is lack of information. Decision makers do not know all that they would like to know about what is going to happen. For situations in which additional information is available at a cost, the procedures of value-of-information analysis can indicate how much additional information is worth and which sources offer the best value for money. The stopping rule for the acquisition of information is, roughly speaking, the point at which the cost of additional information is greater than its expected contribution to the decision being made. Following this rule allows decision makers to optimize their allocation of resources. When an additional (and interested) party is involved, as in informed consent situations, additional constraints arise: that party should, of course, aid the decision maker to analyze how much additional information would be worth. Furthermore, he or she should provide at no cost any available information that could affect the decision maker's actions, thereby eliminating any imbalance of power arising from unequal access to information. Finally, the other party should be willing to invest in the acquisition of new information (of use to the decision maker) up to an amount equal to the net benefit that he or she may gain from the decision maker's choice. These conditions must be fulfilled if the other party wishes to claim to have taken all reasonable steps toward ensuring that the potential consenter has received all the information needed to reach an optimal decision.

Receiving information is, however, no guarantee of being able to use it. The predominant finding of empirical research into judgment under uncertainty is that most intuitions for understanding probabilistic processes are poorly developed.[13] Lacking a grasp of basic statistical concepts, people substitute various rules of thumb. These heuristic devices enable them to produce some response to all judgmental (and decision making) tasks and reasonable responses to most. They can, however, lead to substantial and predictable biases. One example of a heuristic is availability, the device of judging events to be likely to the extent that it is easy to remember or imagine them occurring.[14] Although available events are often relatively likely, they may just be unduly salient. Thus, for example, people tend to overestimate the frequency of lethal events that are disproportionately reported in the news media.[15]

These judgmental difficulties have several implications for the optimality of decision making in informed consent cases. One is that people's intuitions are not to be trusted. It cannot be assumed that information about risks is common knowledge, nor that the obligation to inform people is served simply by depositing technical information on their doorsteps. Rendering information comprehensible requires not only simplifying it and stripping it of technical jargon, but also presenting it in a form that is compatible with people's intuitive ways of thinking. Communication requires an understanding of the recipient as well as of the substance of the message and may require different presentations for different recipients.

A second implication of having unsteady intuitions is that it opens ways to manipulate people's perceptions. People may be reluctant to take a drug that doubles their chance of contracting a dangerous condition (as a side effect), but be less hesitant if told that it raises that chance from 1 in 80,000 to 1 in 40,000. Formally equivalent expressions of risk may not be psychologically equivalent, a fact that may be exploited to aggravate or minimize the concerns evoked by a particular threat.

A third implication is that the efforts needed to convey a particular kind of understanding may be greater than those typically envisioned for a simple informed consent exercise. The decision maker may need to have a feel for what a 1 in 1,000,000 chance of dying (or a one-day loss in life expectancy) means. Yet such intuitions are not created instantaneously.

Finally, although the effects of many biases are quite erratic (sometimes increasing, sometimes decreasing risk perceptions), some are quite consistent. Perhaps the most noticeable of these is the tendency to exaggerate one's personal control over situations, hence one's immunity to risks arising from them.[16] Thus, the rate of consent may be increased simply by letting people overestimate their personal invulnerability. Where such biases are known to exist, the other party bears some responsibility for protecting people from themselves. Conversely, the other party should have some right to counter biases that are unfairly prejudicial to obtaining consent.

Combining All Considerations

Putting all the pieces together in order to reach a decision requires three skills: the ability to select a suitable decision rule, the ability to execute that rule, and the ability to appraise the quality of the product. Careful assemblage of the pieces is of little value if one cannot productively assimilate them. Conversely, recognized inability to assimilate may discourage careful assemblage and contribute to unthoughtful decision making. Not knowing how to decide may precipitate flight from deliberative decision making and retreat to maxim-based choice of action. In informed consent situations, these maxims could include, "Do what I've always done," "Do what others are doing," and "Do what authorities tell me to do."

Such nonanalytic decision rules can, at times, be justified. Because the act of decision making has costs, even economists adopt maxims such as "Always buy the house brands at Safeway; they generally give good value for the money," "Ask Jake about insurance questions; he's thought about them a lot," or "You can trust physicians' advice nowadays; they're so concerned about malpractice that they will exercise about the same degree of caution as you would yourself." These maxims are suboptimal in the sense that a superior result could be had by treating each case separately, but optimal in the sense of one's overall allocation of effort. That a case can be made for reliance on experts, habit, or common practice as a decision rule does not, however, ensure that such reliance will be justified in any particular situation. For a maxim-based choice of action to be treated as an instance of informed consent, the maxim must to a first approximation choose the optimal course of action for the individual involved.

When people make analytical decisions, their behavior can typically be described as following a compensatory rule:[17] the attractiveness of an option increases along with increases in the attractiveness of its positive outcomes and the likelihood that they will be obtained; attractiveness decreases with increases in the unattractiveness and likelihood of its negative outcomes. An appealing feature of compensatory rules is that they are fairly insensitive to computational errors. Thus the difficult mental arithmetic involved in integrating many facts and values need not be too serious a barrier to identifying the optimal option, or one close to it. Indeed, mental arithmetic may pose less of a threat to the execution of decision rules than fear of it poses to their selection. The fear that they cannot "juggle all those things in their head" may lead people prematurely to opt out of analytic decision making and into maxim-based choices. Although they are right to recognize their limited capacity for making precise mental computations, people may exaggerate the threat posed by imprecision.

Implicit in reliance on compensatory rules is the idea that everything has its price. There is no bad that cannot be compensated by some benefit. Such rules underlie so much social commerce that it may be hard to remember that there are other rules, which may flatly rule out certain bads as unacceptable. If, for example, another party offers (what seems to it to be) a good deal, only the most independent-thinking respondent may counter with "there are no deals to be made here; I have a right to an environment without unnecessary risks" or "the idea of hazard pay is immoral; no one should be asked to sacrifice health in order to make a living." Imposing an inappropriate decision rule is inimical to informed consent in this model.

As described here, optimal decision making is a linear and deterministic process, albeit a complex one. One proceeds step by step and eventually adopts the option the decision rule indicates to be best. That depiction is, however, misleading for at least two reasons. One is that strict adherence to this scheme, even in one of its textbook formalizations, carries no assurance of getting it right the first time. Some would, in fact, argue that the signature of an effective decision-making process is that it begs for iteration, repeating the steps in an effort to exploit the insights gained from the first round. Those may include new options meriting analysis, new sources of information worth consulting, or new reflections on value questions. The second inherent limit of analytical decision making is that it creates a model of the decision problem. If that abstraction is inadequate, the recommendation that it produces carries no logical force. Thus the interested party may lead potential consenters through a decision-making scheme pointing to the desirability of the focal option, yet still be told that "it doesn't feel right."

Such disquiet may lead to rejection or repetition of the decision-making process. How defensible it is depends upon the decision makers's degree of insight into the limitations of the process. The empirical evidence leads one to expect under- rather than oversensitivity. As mentioned earlier, considerations that are out of sight often are out of mind. Thus, it is hard for people to realize spontaneously which possible options have not received adequate attention or which consequences have been neglected. A further complication is a marked tendency for people to overestimate the quality of their own knowledge.[18] This overconfidence should lead people to be insufficiently critical of decision analyses and inappropriately willing to accept their conclusions.

Aids to Informed Consent

The preceding section offers a long (yet incomplete) litany of threats to the optimality of decisions.[19] If informed consent is understood to be given whenever the risky option is the optimal choice, then each of these problems is also a threat to informed consent. In order to claim that informed consent has been given, the other party must do all that is reasonable to ensure optimality (or near optimality). Some of the steps that might be taken include the following:

In order to ensure that a full set of options is considered, the other party can develop and present a list showing what other people have done in similar situations and what they might have done had they been more creative or assertive. The length of the list should be commensurate with the resources available for its contemplation. Its members should be sufficiently detailed to be readily comprehended. It should, in particular, include those awkward options whose existence the other party would rather not acknowledge—those expensive concessions that he or she would be willing to make in order to secure consent, but would rather not have as public knowledge. The set should be held open both within each decision-making situation (in order to accommodate those creative solutions that sometimes come with increased familiarization) and across decision makers (in order to exploit others' creativity).

Ensuring proper evaluation of consequences also requires homework on the part of the other party. Again, there is a need for a list of conseqences the potential consenter may wish to consider. Here, too, the list must be sensibly compiled, broad enough to include all potentially pertinent consequences but succinct enough not to inundate the recipient. Although it is up to the decision maker to assign these consequences relative values, it is unfair to leave decision makers to their own devices when the consequences or the comparisons between them are novel. In such cases, the decision maker's immediate response is unlikely to be thoughtful and is likely to be influenced by the way in which the other party has chosen to pose the problem. Here, the other party should offer a variety of perspectives for the decision maker's consideration and attempt to explicate the rationale underlying each. (What kind of person would have these values? How would they be expressed in other decisions?)

Ensuring a proper appraisal of the probability that the consequences will be incurred requires the pertinent information to be available and comprehensible. The availability requirement is met when the other party shares his or her knowledge and acquires reasonably priced additional information. Comprehensibility is harder to achieve. Elimination of jargon is an obvious step. Making information compatible with people's intuitive way of thinking is more difficult—and even dubious—when those intuitions are flawed. In such cases, the information provider's logical choices are between trying to improve or trying to counteract those intuitions (i.e., between education and manipulation of the recipient).

At the stage of combining all these considerations in order to reach a decision, the other party has both easy and difficult tasks. The relatively easy form of help is mechanically aggregating the decision maker's values and beliefs according to several possible decision rules in order to identify candidate best options. If all rules point to the same option, the message is clear. If recommendations diverge, then guidance is needed regarding what logic each rule embodies. The difficult task is helping the decision maker to appreciate the limits of the decision-making process, and to assess how far its conclusions can be trusted.

To the extent that the other party can provide such help when needed, the optimality of the resultant decision should increase, along with the extent to which it constitutes informed consent. Providing this guidance requires specific knowledge of the particular decisions to be made and the capabilities of the people making them. The providers can, however, draw upon the general knowledge and techniques developed within behavioral decision theory (and related fields). In doing so, they will act, in effect, as counselors, helping decision makers to reach the decisions in their own best interests.

This effort is prone to partial failure in at least two respects, both of which can be seen in the preceding discussion of opportunities for assistance. One is that it may be technically impossible to provide the needed help. If, for example, the decision maker's intuitions regarding probabilistic processes are seriously flawed, there may be no corrective techniques available or resources to apply those that are (within the confines of the informed consent exercise). The second common kind of failure is nonneutrality. Even if the other party overcomes this temptation not to share his or her knowledge fully with the decision maker, there remains the danger that presentation of the problem will still be anchored in the other party's perspective, reflecting failure to empathize with the decision maker's concerns. The use of multiple and flexible presentations is designed to counter this problem. It has yet to be shown empirically that it can be done in practice.

What happens when (near) optimality cannot be assured? If neutrality is the problem, an obvious possibility is to replace the other party with an actual decision counselor, like a court-appointed lawyer, entrusted with helping the decision maker and empowered to extract information from the other party or purchase it from other sources (within some budget constraint). If comprehension is the problem, then that external counselor might be asked to serve as a surrogate for the decision makers, that is, try to identify the options that they would choose were they equipped with the resources and skills needed to achieve optimality. Done well, this sort of intervention would lead to better decisions at less cost. A moderate intervention of this sort would be to simplify the problem by screening options that the decision maker needn't take seriously (knowing that some might occasionally prove relevant and that the procedure constitutes a denial of information). A more extreme intervention would be a paternalistic normative analysis using the best available data for evaluating consequences and their likelihood and a combination rule drawn from economic theory, without directly consulting the decision maker. Alternatively, the decision counselor could conduct an extended interview designed to extract whatever beliefs and values the decision maker can give; the counselor would then correct clearly erroneous beliefs, fill in unclear values, and apply some sensible decision rule in order to derive a (binding) recommendation.

Is Informed Consent Only a Matter of Optimal Decision Making?

An oblique answer is "no more than life is only a matter of optimal decision making." Considering the limits of the present analysis may suggest some general limits to the value of the decision-making paradigm as a description of and guide to action. Most of this essay has been devoted to the technical obstacles to achieving optimality. What happens when you get it?

The model of optimality used here envisioned a single individual wrestling with a set of options, values, and beliefs. One obvious limit of this scheme is that the potential consenter is not alone. There would be no consent problem if there were not another party with a vested interest in the outcome of the decision-making process. The key side conditions placed on the process were, in fact, responses to that other party. In particular, they tried to redress the decision maker's inferior familiarity with the problem and access to information. Requirements such as full disclosure of all information can neutralize some of that disadvantage. However, in the course of doing so, they remove the element of negotiation that would otherwise characterize the interaction. It is an interaction between unequals, but an interaction nonetheless and one that would produce a qualitatively different relationship between the parties than the sort of stylized interaction described here. Although the natural relationship in these situations may be conflictive and exploited by the other party, it can also produce the sort of creative solutions and mutual commitment that sometimes arise when people search for a compromise path.[20]

Just as it has no place for negotiation, the model used here has no place for other consequences that arise from the decision-making process. For example, although it may be clumsy and suboptimal, decision making by inexperienced individuals is necessary if they are to increase their skill level.[21] One may feel that informed consent decisions are too important to be used for trial-and-error learning or, alternatively, that the short-term loss from grappling ineffectively with important decisions is a legitimate sacrifice for the acquisition of competence that pays off in the long run. If optimality is all, then paternalistic solutions are tempting, as long as there is good reason to believe that some other party is best able to identify the course of action that is in the potential consenter's interest.

The decision-making process can bring about other changes that may also justify some sacrifice in expected optimality. Even when it cannot produce any general improvement in decision-making skills (e.g., because the feedback needed for learning is absent), hands-on grappling may still provide valuable familiarization with the facts of the particular problem. That knowledge may be necessary if the decision maker is to monitor the fidelity with which the consent agreement is implemented or to be alert to any risks that are realized. A final realm of changes that may be worth considering are those occurring within the decision makers themselves. People are different for making decisions in their own behalf, with increased pride and self-confidence.

On the other hand, these nonoptimization goals may be unobtainable or negligible (compared with the stakes riding on the informed consent decision). Citing them may serve merely to deprive the potential consenter of needed help in an unequal relationship with the other party.

Optimality is not everything. But it is a lot. Everything reasonable should be done to help people identify the course of action that best serves their interests, whether or not that decision will be interpreted by some other party as constituting informed consent.

Concluding Summary

In making a decision, one makes a choice between alternative courses of action. If the chosen option involves some risk, then another party may claim that the choice represents informed consent to the risk that the option entails. These risks have special status for the other party, but not for the individual making the choice, whose goal it is to choose the best option, considering all possible consequences. In order to interpret the choice of the risky option as an act of informed consent, the other party must be able to claim that it has constituted the individual's optimal choice. To make this claim, the other party must help the individual to overcome some difficulties that are inherent in intuitive decision making and avoid creating new difficulties arising from the unequal power of the two parties.

Notes

1. Actions taken out of habit might still be considered to be decisions if they were deliberatively chosen among a set of alternatives at some previous time.

2. Here would be included the good things associated with risks, such as the thrill or prestige that they bring.

3. Convenient entry points to this literature may be found in W. Edwards, "A Theory of Decision Making," *Psychological Bulletin* 51 (1954): 381–417; W. Edwards, "Behavioral Decision Theory," *Annual Review of Psychology* 12 (1961): 473–98; B. Fischhoff, S. Lichtenstein, P. Slovic, S. Derby, and R. Keeney, *Acceptable Risk* (New York: Cambridge University Press, 1981); R. A. Howard, "The Foundation of Decision Analysis," *IEEE Transaction, on Systems, Science and Cybernetics*, SSC-4, no. 3 (1969): 393–401; R. L. Keeney and H. Raiffa, *Decisions with Multiple Objectives* (New York: Wiley, 1976); H. Raiffa, *Decision Analysis* (Reading, Mass.: Addison-Wesley, 1968); E. Stokey and R. Zeckhauser, *A Primer for Policy Analysis* (New York: Norton, 1978).

4. Entry points to this literature include Fischhoff et al., *Acceptable Risk*; R. Hogarth, *Judgment and Choice: The Psychology of Decision* (Chichester, England: Wiley, 1980); D. Kahneman, P. Slovic, and A. Tversky, eds., *Judgment Under Uncertainty: Heuristics and Biases* (New York: Cambridge University Press, 1982); R. E. Nisbett and L. Ross, *Human Inference: Strategies and Shortcomings of Social Judgment* (Englewood Cliffs, N.J.: Prentice-Hall, 1980).

5. Discussions of the sensitivity of decisions to errors may be found in B. Fischhoff, "Clinical Decision Analysis," *Operations Research* 28 (1980): 28–43; H. Raiffa, *Decision Analysis*; D. von Winterfeldt and W. Edwards, "Costs and Payoffs in Perceptual Research," *Psychological Bulletin* 91 (1982): 609–22.

6. E.g., C. E. Rosen, "Sign Away Pressures," *Social Work* 25 (1976): 284–87, who found a reduction from 100 percent to 20 percent in clients' willingness to release information from their personal files, once they were told that refusal was possible.

7. Further discussion of these issues and references to research may be found in B. Fischhoff, P. Slovic, and S. Lichtenstein, "Knowing What You Want: Measuring Labile Values," in *Cognitive Processes in Choice and Decision Behavior*, edited by T. Wallsten

(Hillsdale, N.J.: Erlbaum, 1980); H. Schumann and S. Presser, *Context Effects in Survey Research* (New York: Academic Press, 1981); C. Turner, E. Martin, and T. DeMaio, *Survey Measure of Subjective Phenomena* (Washington, D.C.: National Academy of Sciences, 1982).

8. A. Brooks and B. A. Bailar, *An Error Profile: Employment as Measured by the Current Population Survey* (Washington, D.C.: U.S. Department of Commerce, 1979).

9. Indeed, it may not even be necessary if one considers the possibility that all people may not interpret a given situation in the same way. Standardization, like so many other aspects of interviewing, is defined from the perspective of the interviewer.

10. The classic reference is S. L. Payne, *The Art of Asking Questions* (Princeton, N.J.: Princeton University Press, 1952).

11. A common tactic among politicians' personal pollsters is to ask a variety of seemingly equivalent questions and report only the ones that express the greatest support for their candidate. In the Pinto trial, Ford Motor Company is reported to have commissioned a poll to determine what defense position would be viewed most favorably by samples of subjects drawn to be similar to the trial jury.

12. It would seem as though the moral strictures governing manslaughter or suicide would be better applied here than those regarding informed consent.

13. Most of the standard references for this field may be found in Kahneman, Slovic, and Tversky, *Judgment Under Uncertainty*.

14. A. Tversky and D. Kahneman, "Availability: A Heuristic for Judging Frequency and Probability," *Cognitive Psychology* 5 (1973): 207-32.

15. S. Lichtenstein, P. Slovic, B. Fischhoff, M. Layman, and B. Combs, "Judged Frequency of Lethal Events," *Journal of Experimental Psychology: Human Learning and Memory* 4 (1978): 551-78.

16. E. Langer, "Illusion of Control," *Journal of Personality and Social Psychology* 32 (1975): 311-28. O. Svenson, "Are We All Less Risky and More Skillful Than Our Fellow Drivers?" *Acta Psychologica* 47 (1981): 143-48. This particular bias could be viewed as a case of emotion clouding reason, or wishful thinking. However, it could also be explained by pointing to improper information processing due to the lack of statistical intuitions. It is a bias of behavioral decision theory to look first for nonmotivational explanations of poor performance. This predisposition reflects both the "cognitive backlash" against psychoanalytic accounts and the hope of finding ameliorative measures.

17. This need not mean that they are actually following such rules. The robustness of compensatory rules discussed further on in the text enables them to describe approximately quite varied kinds of behavior. See B. Fischhoff, B. Goitein, and Z. Shapira, "SEU: The Story of a Model," *Journal of the American Society of Information Sciences* 32 (1981): 391-413 (and references therein).

18. See S. Lichtenstein, B. Fischhoff, and L. Phillips, "Calibration of Probabilities: State of the Art to 1980," in Kahneman, Slovic, and Tversky, *Judgment Under Uncertainty*.

19. Further verses in this litany may be found in Kahneman, Slovic, and Tversky, *Judgment Under Uncertainty*; or Fischhoff, Lichtenstein, Slovic, Derby, and Keeney, *Acceptable Risk*.

20. The theory of games addresses some of these questions by considering optimality in the context of n-person situations. It doe not, however, deal with topics like the creation of new options or the quality of the personal relationship between the parties.

21. R. M. Hogarth, "Beyond Discrete Biases: Functional and Dysfunctional Aspects of Judgmental Heuristics," *Psychological Bulletin* 90 (1981): 197-217.

9

The Role of Consent in Managing Airborne Health Risks to Workers

Michael Baram

Airborne Health Hazards in the Workplace

Workers are exposed to a wide variety of health hazards from airborne contaminants in the workplace.[1] The contaminants include asbestos, lead, beryllium, vinyl chloride, coal dust, other chemicals and trace metals, and various biological and radioactive substances. The health risks to workers depend on the toxicity and other hazardous attributes of the contaminants; the incidence and magnitude of exposure; the work habits, life-style (e.g., smoking), and biological attributes of the exposed worker; and the presence of other relevant promotional or inhibitory factors.[2]

Over time (the latency period), such health risks accrue and can become manifest as serious and even fatal occupational diseases. Cancer, respiratory impairment, neurotoxicological illness, and reproductive disfunction are among the occupational diseases that now befall workers exposed to various airborne contaminants on the job.

Employers use a variety of engineering and health-based strategies to reduce or eliminate such health risks.[3] Some of these strategies are mandated: they are based on regulatory requirements (e.g., OSHA regulations) or on contractual agreements (e.g., provisions of collective bargaining agreements between union and management).

Other strategies are applied by employers at their discretion in order to reduce potential company liability that may arise later from legal determinations, as a result of claims filed by workers in court or under workers' compensation or other legally established insurance or indemnification systems (such as the black-lung program).[4] These discretionary strategies may also be designed to prevent other economic losses or increased costs, such as losses from reduced worker productivity caused by job-related illness and union demands for increased wages and benefits to offset workplace health hazards. Thus, corporations respond, to varying degrees, to worker health risks when these bear on corporate profitability, regardless of any applicable laws.

The mandatory and discretionary strategies of employers involve engineering controls applied to the industrial processes (e.g., ventilation and containment systems); the use of substitute products and other changes in the industrial processes; initial health examinations and later ongoing medical surveillance of workers (biological monitoring); ongoing monitoring of the workplace atmosphere to identify and remove workers at risk prior to the manifestation of illness (ambient monitoring); and the use of special clothing, respirators, and other self-protective measures by workers.[5]

Technical uncertainty prevails. How much health risk is job related? How much is exacerbated by worker smoking or genetic characteristics? Economic implications are significant to management and shareholders and to unions and individual workers. Industrial health and medical professionals, engineers, and attorneys are employed at considerable cost and are sometimes caught in conflicts between their professional responsibilities to workers and their duties to their employers. Unions and workers seek company-held information to better protect their interests, but companies often claim that such information is proprietary and assert their right to protect the value of their property interest in the information by keeping it secret. Litigation flourishes, as reflected in increasing asbestos disease claims made against employers and their insurers.[6]

Thus, the task of managing airborne health risks to workers is complex, costly, and controversial. In this dynamic and troubled framework, what is the role of consent?

The Issue of Consent

To grapple with this question, one must consider several different uses of the concept of consent. First, "consent" is commonly used to denote the act taken "to agree or . . . give assent to something proposed" or to denote the "agreement, approval, acquiescence" itself. Thus, consent commonly indicates a "unity of opinion which involves the presence of two or more persons."[7]

Second, consent is one of the key elements of a contractual relationship, a contract being, in general, "an agreement upon sufficient consideration to do, or refrain from doing, a particular lawful thing." The consent may be express and affirmative in nature, or implied or passive in the sense that acquiescence, as a form of consent, can sometimes be construed from silence.[8]

Third, "consent" and "the contractual relationship" have been frequently adopted in law and in philosophy as concepts that can be used to guarantee the protection of individual interests and, in a larger sense, can contribute to the promotion of fairness, justice, and equality in society. Thus the principle of "informed, voluntary consent" has been developed in the common law to ensure that medical patients maintain some control over the experimental and therapeutic measures (and risks) to be applied to them by the medical professionals they hire.[9]

However, the concept of consent in the common law is not based solely on the contractual relationships that may exist between parties. Consent has served as a key element in the evolution of tort law—that loosely structured field of common law that is used to compensate victims for various harms in situations or relationships that fall outside the boundaries of contract law and criminal law.

Tort law provides, for example, that "a person has a duty not to permit his activities to create an undue risk of harm to any protected interest in another," but an exception is generally recognized: the creator of the risk "may not be liable to the other . . . because of the consent . . . of the other." Thus, consent of the victim is a common defense in tort cases: "A person of full capacity who, with full knowledge of the facts, freely consents . . . to the conduct of another is not entitled to recover for harm caused to his interests by such conduct."[10]

These historic tort law principles, and the role of consent in tort law, reflect the underlying tension in this field of law. In determining if compensation is a proper remedy for harms arising to one party from the actions of another, the court has to undertake an approach that somehow balances, in each case, the conflicting interests of the parties, both the plaintiff's "interest in security" and the defendant's interest in "freedom of action."[11]

Philosophers have also drawn on the concept of consent in propounding theories of equality and justice for western, industrial society. Rawls, for example, puts forward a theory of justice based on a hypothetical social contract, which identifies just arrangements as those individuals would have consented to in an initial situation of fairness.[12] In this view, concepts of contract and consent underlie the notion of what constitutes a just state in which individuals are treated and respected as equals.

But as another philosopher, Patrick Derr, has pointed out, the concept of consent can also be appealed to to justify certain inequalities: for example, to support the current differences in the degree of protection afforded workers and the general public from the same hazardous substances. The consent argument runs: "an allocation of risks is just if, and only if, it has the consent of those upon whom the risks are imposed." This argument has been used to support less stringent standards for worker protection.[13]

Thus, consent is a concept at work in our legal and value systems and as such helps to shape and define how we manage occupational health risks to workers. But the concept remains elusive, qualitative (unmeasurable), and troublesome to philosophers and policy analysts alike. As Derr notes, "There are deep and heated disagreements over the question of what precisely constitutes morally adequate consent . . ."[14] Nevertheless, there is consensus in law and philosophy that two attributes of consent are of most importance in defining its moral and legal adequacy in any particular circumstance: the extent to which the consent involved is "voluntary" and "informed."[15]

The voluntary measure of consent means that the consenting subject is (a) competent to give consent and (b) free of coercion. In the American workplace, where virtually all workers can be considered competent (given legal restrictions on hiring minors and the protection of guardians for the mentally retarded) and are capable of leaving their employment at will, the voluntary criterion can generally be reduced to one issue — what Derr has called "structural coercion." This is coercion "rooted in the subject's situation [such as] poverty . . . to the extent that it undercuts a subject's ability to decline a risky job, high unemployment, or possession of relatively few marketable job skills."[16]

The informed criterion of adequate consent, as described by Derr, raises the question: "What information is required by a subject in order for that subject's consent to be considered adequately informed?"[17] From a practical legal and ethical standpoint, it is the more appropriate focus of concern for at least two reasons: (1) "structural coercion" is essentially a national policy question that transcends any of the applicable legal frameworks; and (2) assuming voluntariness, proper disclosure of health risks ensures "respect [for] the autonomy of individuals in making basic life decisions [makes] legitimate the distribution of risk, and [serves to] enhance the efficiency of efforts to reduce risk."[18]

Given this background, one can now assess the role of consent in managing airborne health risks to workers in three legal frameworks available for protecting worker health: the common law; workers' compensation law; and federal regulation under the Occupational Safety and Health Act.[19] Certain useful questions to ask are:

(1) Is consent provided for?
(2) By what process is consent reached?
(3) How is consent guaranteed?
(4) How voluntary is consent?

(5) How informed is consent?
(6) What are the major obstacles to improving the voluntary and informed qualities of consent?
(7) How useful is consent as a factor in managing occupational disease risks?

These questions are addressed in the following sections.

Common Law

The common law has dealt with the problem of worker exposure to health risks through the application of general principles that define the master-servant (or employer-employee) relationship in terms of rights and duties. Although these rights and duties could be construed as growing out of an express or implied contractual relationship, the courts chose to ground them in the tort or negligence principle that a property owner must not negligently expose to danger persons who come upon the premises by invitation or to transact business. The courts then modified this general duty in particular cases involving express and implied contracts of employment, so that masters would not be liable for injuries to servants who knowingly consented to or assumed those obvious risks that caused the injuries.[20] However, the enactment of worker compensation laws, subsequently discussed, has virtually eliminated employee rights to bring tort actions against an employer for job-related injury. Therefore, discussion of tort law is provided to illuminate how the tort system would function if worker tort rights are restored in the future.

Little case law deals with the workplace health risk of occupational disease, which is the main concern of this inquiry; most of the cases deal with sudden accidents or safety problems. Nevertheless, the early principles are broad and remain of potential applicability in tort litigation by occupationally diseased workers today. For example:

> A master who carries on an extraordinarily dangerous undertaking, such as the generation . . . of electricity, is bound to know the character and extent of the danger, and to notify the same to the servant specifically and unequivocally, so clearly as to be understood . . . The absence of knowledge is no exculpation; the servant is not required to know latent but only patent defects, and he has the right to assume superior knowledge in his employer, to rely on his prudence and judgment, and to believe that he will not, unnecessarily, expose the employee to danger.[21]

To strengthen this principle, the courts have generally refused to uphold employment agreements that exempt the employer from liability for negligence toward his employees, finding such agreements contrary to public policy and void.[22]

The "assumption of risk" defense by an employer has often been accepted, but has also often been rejected or limited by the courts. The "assumption of risk" argument is based, in large part, on employee consent to risk. The common-law decisions reveal two by now familiar restrictions on the defense: the employer must show (1) that the employee knew and understood the risk he was incurring (informed consent); and (2) that the employee's choice to incur it was free (voluntary consent).[23]

Some general answers to our questions on the role of consent in the common-law framework for protecting workers from occupational disease will now be attempted.[24]

(1) Is consent provided for?
At common law, one's voluntary status as an employee manifests consent. However, the permissible scope of the consent will be limited: it does not include a worker's consent to health risks latent and unknown or latent and undisclosed (by the employer) that lie beyond his normal capacity for detection; nor does it include a worker's consent to hold the employer harmless for negligent conduct by the employer.

(2) By what process is consent reached?
The common law does not dictate any specific process; it assumes that one's employed status reflects the reaching of consent, but bounded in scope, as noted above, even if consent is express in an employment contract. The courts usually will not accept consent when it has been coerced.

(3) How is consent guaranteed?
The common law indirectly guarantees an employee's right to consent (by restricting use of the consent-to-risk argument as a defense in suits against an employer) and thereby provides an incentive for employers to disclose risks and seek consent explicitly from employees.

(4) How voluntary is consent?
At common law, consent has been dealt with mainly as an integral part of the employer's defense that the worker consented to or assumed the risk. The courts have often accepted this argument, but rejected it in many cases where it has been shown that the consent was not voluntary or free of duress or other forms of coercion. Thus, courts have found coercion and no consent in cases where an employee was given a choice between two "evils."

Of course, the scenario of most concern to this inquiry involves an employer's use of economic pressure (threatened firing) to secure an employee's consent to work under certain risk conditions. Prosser has described the controversies in the United States that arose when most courts held that a "risk was assumed [consented to] even when a worker acted under a direct command carrying an express or implied threat of discharge for disobedience." He concludes, however, that the judicial acceptance of this form of coerced consent as a defense and the hardships it generated were adroitly sidestepped by state legislation establishing an alternative route for worker remedy, namely, the workers' compensation system.[25]

We now know that the workers' compensation system is inadequate in many respects, such as in the compensation afforded workers and the evidentiary difficulties of meeting the requirement of proving the job relatedness of an occupational disease. Thus, many workers are looking again to tort law for more adequate remedies, and judicial acceptance of consent under economic coercion may once again become a major problem.

(5) How informed is consent?

Prosser has concluded that "under ordinary circumstances the plaintiff will not be taken to assume any risk of . . . activities or conditions of which he is ignorant." Thus, the defensive use of consent in the assumption-of-risk argument faces the limitation that the consent must be knowledgeable or informed. For airborne health risks in the workplace—invisible, often without odor, and not perceptible to ordinary persons—the duty is clearly on the employer to inform the worker, if the employer is to avoid liability in the form of compensatory damages; and as developed in asbestos-related litigation, additional liability in the form of punitive damage awards.

By requiring employer disclosure of material information regarding personal health risks as a prerequisite for the employer's later use of consent to risk as an effective defense in tort litigation, the courts have sought both to promote the employee's individual autonomy and to establish an equitable and rational basis for employer-employee relationships under conditions of risk. The former function is carried out by judicial insistence that the person at risk be the ultimate decision maker regarding the risk in question; the latter function is realized by judicial insistence that employers have a duty to learn of the risks and disclose all material information to the employee before an agreement to assume risk will have legal effect.

(6) What are the major obstacles to improving the voluntary and informed qualities of consent?

Probably the major obstacle involves uncertainty as to what constitutes risk. Risks can be defined by any of several analytic methods and chosen assumptions. For chemical risks, for example, a critical issue is whether one should assume a risk threshold or a linear dose-response relationship, or some other model. Another issue pertains to analytic method: should a risk-benefit approach or other (e.g., zero risk) approach be used? Further, is extrapolation from ambient levels of contaminants a valid basis for finding risk, or is biological monitoring or medical evidence needed (such as routine testing of individual workers' blood-lead levels)?

Employers also object to the consent doctrine's disclosure requirements. Disclosure of alleged trade secrets or proprietary data may lead to a loss of company property; other information disclosed (worker medical test results, etc.) may be self-incriminating and lead to liability. These are but a few of the many obstacles.

(7) How useful is consent as a factor in managing occupational disease risks?

Consent as developed in the common law is a dynamic factor of considerable potential for reducing worker risk, provided the common-law system effectively functions and imposes liability in practice. Probably because of this potential, industry has sought to limit the applicability of the common law to occupational disease risks and has lobbied for ineffective worker compensation and federal regulatory programs to deal with occupational disease. Once established, these programs have worked to exclude or limit use of the common-law system, without effectively acting on occupational disease risks.

Workers' Compensation

Each of the fifty states has a workers' compensation system in effect. Since the early years of this century, these systems have worked to provide compensation to injured and disabled workers, and in recent years, they have been expanded to cover occupationally diseased workers, provided they are able to prove that their disease is job related.[26]

Support for workers' compensation has come from two sources: (1) a beneficent societal concern for providing injured and diseased workers with compensation without recourse to tort law (with its tougher evidentiary requirements for the worker-claimant to meet and with its assumption-of-risk defense effectively used by employer-defendants); and (2) industrial support for the structured workers' compensation system, so long as it provides the exclusive remedy to the worker and is therefore in lieu of other remedies available under tort law with its many vagaries (such as high jury awards).

Because of many inadequacies in workers' compensation, such as the low level of compensation provided in most states, courts in recent years have permitted tort actions by worker against employer in certain cases, in addition to worker claims under workers' compensation law, thereby eroding the principle that the workers' compensation remedy excludes tort actions involving the same parties and injury.[27]

The claimant in a workers' compensation proceeding must usually prove injury or disease (diagnosis), its job relatedness (causation), and the extent of the subsequent disability (prognosis) to secure the compensation award from the cognizant state board and to withstand any subsequent legal challenges in state court. The defendant insurer (of the claimant's employer) can choose to defend in these proceedings and in some cases does so. An adverse decision may be further appealed in state court. In the usual occupational disease claim case, however, the insurer and the claimant will settle the claim privately.

In this structured compensation system, many of the issues that would be important in tort litigation are of no consequence. Whether or not consent to risk was reached, and was informed and voluntary, are issues not raised in workers' compensation proceedings. Thus, assumption of health risk by a worker, fully informed and voluntary, is not available as a defense to the insurer representing the employer's interests.

Therefore, the questions set forth earlier can readily be answered for the workers' compensation system, as follows:

(1) Is consent provided for? No.
(2) By what process is consent reached? Not relevant.
(3) How is consent guaranteed? Not guaranteed.
(4 and 5) How voluntary and informed is consent? Not relevant.
(6) What are the major obstacles to improving the quality of consent? Consent issue is not relevant in proceedings before workers' compensation boards under present law. Legislative action in fifty states would be needed to effect change.
(7) How useful is consent in managing occupational health risks? The concept of consent is of no consequence in the context of workers' compensation.

OSHA

The Occupational Safety and Health Act of 1970 (OSH Act) is administered by the Occupational Safety and Health Administration (OSHA), a division of the U.S. Department of Labor. OSHA sets workplace health standards and inspects and enforces for compliance.[28] Several of its standards deal with workplace air pollutants, such as lead.

In enacting this statute to protect workers through preventative regulation Congress did not intend to diminish worker opportunities for compensation under the state workers' compensation systems, nor diminish worker rights to sue in tort. Thus, the OSH Act does not preempt private rights of action by workers against employers and their insurers and is silent on the issue of consent to health risk.

Accordingly, OSHA, in administering the act, has not directly addressed the consent issue in its extensive regulations. To do so would be ultra vires: beyond the agency's authority. Nevertheless OSHA has enacted many regulations that are consistent with the common law concept of voluntary, informed consent and indeed seem to be based on or sustained by it.

For example, OSHA has by regulation required worker access to medical and other health-risk records kept by employers;[29] has established a "self-help" right to refuse hazardous work when a worker is confronted by a condition on the job that places him in "imminent danger" of serious bodily harm;[30] and has recently proposed labeling requirements for hazardous materials in the workplace.

These and other OSHA enactments promote worker consent to health risk on a more fully informed basis.[31] Further, by establishing a self-help right in the case of "imminent danger" of serious bodily harm (but not yet for "danger of chronic disease"), OSHA has eroded somewhat the "structural coercion" that constituted a major obstacle to the exercise of truly voluntary consent to health risk by a worker. Therefore, OSHA actions have reinforced the worker's informed and voluntary consent rights somewhat within the OSHA framework.

On the other hand, these OSHA actions may make it more difficult for the worker seeking to impose tort liability against the employer or a third party to argue lack of material information regarding risk and may make it easier for the employer or third party to argue that the worker assumed the risk on an informed basis; or at least had the right to secure the risk information needed. This is speculation until further developments clarify the implications of the OSHA actions discussed above.

To answer the questions set forth above in the OSHA context:

(1) Is consent provided for? No, but certain elements of consent are affirmed and indeed reinforced by OSHA regulations.
(2) By what process is consent reached? Not addressed.
(3) How is consent guaranteed? Not guaranteed.
(4 and 5) How voluntary and informed is consent? Consent is made more meaningful by OSHA regulations that support and strengthen worker rights to risk information and self-help.
(6) What are the major obstacles to improving the quality of consent? Key obstacles include industrial and executive office (OMB) pressures on the agency to rescind or weaken "access to information" and labeling regulations, and OSHA's laggard enforcement efforts since 1980.
(7) How useful is consent in managing occupational health risks? In the OSHA context, regulations promoting access to information, labeling, self-help, and other aspects of consent impose new pressures on industry employers to reduce health risks or otherwise face legal and economic consequences. In the common law context, the effect of OSHA regulations may be to diminish workers' opportunities for remedies under tort law.

To sum up:

(1) Consent is a vital aspect of the common law for managing health risks, since it has a continuing forcing effect on employers to reduce risks and disclose information about risks. But this forcing effect is diminished or blunted by:

- uncertainty as to the extent to which an employer has an affirmative duty to identify and measure workplace health risks;
- uncertainty as to what constitutes an unreasonable health risk;
- uncertainty as to what disclosure is needed to bring about properly informed consent by employees;
- the problem of "structural coercion," which cannot be readily solved by the courts or agencies, other than through the provision of new guarantees to workers to ensure their ability at self-help efforts; and
- the judicial acceptance of assumption-of-risk defense arguments in certain cases devoid of adequate informed and voluntary consent.

(2) Consent and assumption-of-risk concepts have been deliberately excluded from the workers' compensation systems of the fifty states, and consent therefore has no discernible role there.

(3) Consent is not an explicit OSHA concern, but the agency has adopted the "informed" and "voluntary" elements of meaningful consent in its enactments, and one can say that the consent concept has been strengthened by OSHA regulations on information access, labeling, self-help, and other matters. Political and economic factors may change the agency's direction on these matters, however, and OSHA enforcement has become a real problem.

Notes

This paper was derived, in part, from research conducted under NSF-EVIST grant number ISP8114738.

1. See Nicholas Ashford, *Crisis in the Workplace* (Cambridge, Mass.: MIT Press, 1976).
2. See Richard Doll and Richard Peto, "The Causes of Cancer: Quantitative Assessments of Avoidable Risks of Cancer in the United States Today," *Journal of the National Cancer Institute* 66 (1981): 1191–308.
3. See Audrey Freeman, *Industrial Response to Health Risk* (New York: Conference Board, 1981).
4. The federal black-lung program is designed to compensate those workers in coal mining who develop pulmonary problems that are job related. The costs of compensation, borne in part by industry, are expected to have a deterrent effect on industrial practices that lead to such worker risks.
5. Based on research projects currently supervised by the author. See author for further details.
6. See Note 5.
7. *Ballantine's Law Dictionary* (Rochester, N.Y.: Cooperative Publishing Co., 1969), p. 249.

8. Ibid.
9. See William Prosser, *Law of Torts* (St. Paul, Minn.: West Publishing Co., 1971), Chap. 11.
10. Ibid.
11. See William Seavey, "Principles of Torts," *Harvard Law Review* 56 (1972).
12. John Rawls, *A Theory of Justice* (Cambridge, Mass.: Harvard University Press, 1971). Also see Ronald Dworkin, *Taking Rights Seriously* (Cambridge, Mass.: Harvard University Press, 1978), Chap. 6.
13. Patrick Derr et al., "Worker/Public Protection: The Double Standard," *Environment* 23, no. 9 (1981).
14. Patrick Derr, quoted in Mary Melville, "Risks on the Job: The Workers' Right to Know," *Environment* 23, no. 9 (1981).
15. Ibid. ". . . there is also a broad and stable consensus . . . that such consent must be free and informed" (p. 15).
16. Ibid.
17. Ibid.
18. "Occupational Health Risks and the Worker's Right to Know," *Yale Law Journal* 90 (1981): 1792.
19. A fourth framework should be noted, although it is not assessed here: collective bargaining between union and management. This framework is discussed by Mark MacCarthy in Chapter 10 of this volume, entitled "Reform of Occupational Safety and Health Policy."
20. Henry Buswell, *Law of Personal Injuries* (Boston: Little, Brown, 1899), Chap. VII.
21. Ibid., p. 403.
22. Prosser, *Law of Torts*.
23. Ibid.
24. It should be noted that the common law on master-servant issues, including consent, is contradictory and extensive, and not easily reducible to conclusive findings. A treatise would be required to deal with all its variations, and the general answers to the questions in the text should therefore not be regarded as complete.
25. See Prosser, *Law of Torts*.
26. See *Annual Survey of Workers' Compensation*, U.S. Chamber of Commerce, 1981.
27. See generally, Arthur Larsen, *Workers' Compensation Law* (New York: Matthew Bender Publishing Co., 1982) and other treatises such as *Massachusetts Practice*, vol. 29 (St. Paul, Minn.: West Publishing Co., 1982).
28. OSH Act, 29 U.S.C. secs. 651-78.
29. 29 CFR sec. 1910.20.
30. 29 CFR sec. 1977.12 (b)(2). See *Whirlpool Corp. v. Marshall*, 445 U.S. 1 (1980). No regulations or case law address the "imminent hazard" of occupational disease.
31. E.g., OSHA's lead standard provides for medical removal, job retention, and salary during removal.

PART THREE

Pollution and Policy

10

Reform of Occupational Safety and Health Policy

Mark MacCarthy

Introduction

In 1970, Congress passed the Occupational Safety and Health Act to ensure "the greatest protection of the safety or health of the affected employees."[1] Congress found that in the absence of a strong government role, safety and health conditions in the nation's workplaces had deteriorated and that the private market could not be relied on to protect workers. In 1971, the Occupational Safety and Health Administration (OSHA) was established to promulgate and enforce regulations in accordance with this legislative mandate.

Over the past decade, OSHA has been embroiled in heated controversy. Critics have charged that OSHA has overregulated, that its standards and enforcement have been nit-picking, intrusive, and too expensive for industry in a time of lagging productivity and economic decline. As a result of this criticism, many, if not most, current discussions of occupational safety and health policy focus on how regulation can substantially achieve its goal at lower cost. This efficiency perspective is sensible and needed, but it has dominated the discussion to the exclusion of a more basic, but more controversial, question, namely, how much should we as a nation do to control workplace threats to safety and health?

This policy choice is quite different conceptually and practically from private decisions concerning risk. An individual worker faces the question of whether a job poses unacceptable risks. As citizens we face the policy question of whether we are doing enough to control workplace hazards. Despite its place in the prevalent cost-benefit approach to health and safety regulation, the individualist perspective is only marginally relevant to the policy decision we face as citizens. This is in part because the current preferences of workers for safety and health on the job are strongly conditioned by the current recession, by lack of knowledge and power, and by an unfair distribution of income. There are also deeper questions concerning whether the simple aggregation of private preferences revealed in the market transactions of private life should ever override politically based social decisions on public values such as those embodied in the OSH Act.[2]

My premise in this chapter is that we are doing too little to control workplace hazards and that current institutional arrangements must be reformed and supplemented if we are to do anything like an adequate job. A partial justification of this premise is provided by available occupational safety and health statistics. While these statistics are not completely accurate measures of safety and health conditions in the nation's workplaces, they suggest that both the absolute number and the rate of workplace injuries and illnesses are unacceptably high. The Bureau of Labor Statistics estimates that, in 1981, 4,370 workers were killed on the job, 5,404,400 work-related injuries and illnesses occurred, and one out of every 13 workers was injured or made ill on the job.[3] The National Safety Council puts the number of work-related deaths in 1980 at 13,000.[4] These estimates do not include longer-term occupationally related illnesses such as cancer. The National Institute for Occupational Safety and Health (NIOSH) has estimated that these longer-term cases may amount to as many as 100,000 per year.[5] Other reported estimates of the number of fatalities from occupational illnesses range from 10,000 to over 20,000 per year.[6]

Official occupational safety and health statistics do not provide any basis for the belief that the situation has dramatically improved over time. Indeed, they reveal little discernible trend except the predictable variation due to the business cycle.[7] As a result, the purely economic burden imposed by workplace injuries and illnesses remains staggeringly large. The National Safety Council estimates that the annual cost of these injuries and illnesses is approximately 1 percent of the Gross National Product, or about $30 billion in today's economy.[8]

Relative to this enormous economic burden, capital investment for safety and health equipment is insignificant. Moreover, it has declined during the 1970s both absolutely and as a proportion of all new capital expenditures. In constant 1972 dollars, capital expenditures on workplace safety and health were $3.30 billion in 1972, but only $2.63 billion in 1981. This represents a decline from 2.7 percent of all plant and equipment investment in 1972 to 1.5 percent in 1981.[9] Far from imposing an enormous regulatory burden, OSHA has promulgated only 23 health standards in the ten years of its existence.

In view of this lackluster performance, the charge of overregulation seems to me to be misplaced. A more accurate criticism is that OSHA has been ineffective in directing more resources to worker protection. In the face of this apparent regulatory failure, what can be done to increase worker protection? An initially appealing but ultimately unsatisfactory approach would be to rely more heavily on the court system to provide greater worker protection. I discuss this possibility briefly in the second section of this chapter.

A second approach would incorporate the objectives of occupational safety and health regulation, and other social regulation, within the framework of a coherent, overall industrial policy. This would recognize industry's valid interest in avoiding overlapping and duplicative regulations. It could also involve providing federal assistance for firms investing in equipment to control workplace hazards. I discuss this approach in the third section.

An alternative approach calls for greater involvement of workers and managers in negotiating protection on both a national and local level. The principle of involving workers in their own protection either directly through their local union chapters or indirectly through their national unions is an appealing social ideal. These mechanisms to reduce workplace risk are desirable in principle because people should not only be protected against workplace risk, but should also have meaningful control over how that risk is reduced. At the national level one way to accomplish this goal of greater worker control over workplace risk would be to provide for more bargaining and consensus building in the standards development process. I discuss this approach briefly in the fourth section.

Centralized policies to protect workers need not only reform from the top down, but also supplementation at the local level. The most promising mechanism for increasing local control over workplace risks is greater use of the collective bargaining process to enforce federal standards and to establish worker-management safety and health committees. I discuss these possibilities in the final section.

The Court System

One appeal of greater reliance on the court system is that the courts provide a theoretically neutral forum where wronged parties can bring complaints based upon traditional notions of fairness, due process, reasonable care, and so forth. This process acknowledges principles behind the formal laws and rules passed by the rule-making bodies of public life, namely, the common law tradition's implicit assignment of rights and duties. The regulatory framework with its reliance on formal rules may in fact discourage efforts to increase safety in the workplace by encouraging both workers and management to adopt a "work to rule" mentality. Responsibilities are limited to compliance with explicit rules, and commonsense precautions that might have been reasonable demanded, but are not required by formal rules, are not taken.[10]

Can appeal to the courts provide a supplement to the market and to regulation that reinforces these notions of responsibility and reasonable care and so increases the efforts made to protect workers? I do not think much can be expected from greater reliance on individual responsibility. The regulatory approach may to some degree undermine a sense of duty, and regulators ought to be more concerned about this than they have been. The effectiveness of a traditional sense of responsibility for providing worker safety and health is open to question. Too often it is undermined by the natural imperatives of a competitive business environment. One of the most corrosive effects of a market system is to replace the complex common law system of rights and duties with the single motive of profit maximization. Indeed, the normal emphasis on pursuit of profit can tend to undermine the pursuit of goals mandated by a traditional sense of duty and responsibility to the community.

While the notion of personal responsibility and common law traditions of due care and reasonable precaution can in principle motivate greater efforts for workplace safety, increased reliance upon the courts will not necessarily foster and encourage these sentiments. Rather, workplace safety rules should be flexible enough so that they supplement and focus these traditional ideas, but do not override them.

Another argument in favor of greater reliance upon the court system is that it can give industry an enormous incentive to provide safer workplaces. This incentive is considerably greater than that contained in the current workers' compensation system, which limits workers' rights to sue firms for negligence in return for "no-fault" compensation for workplace injuries and illnesses. Under the current compensation system workers receive a maximum of two-thirds of their lost income plus medical expenses. The present discounted value of these awards for an occupational illness that may manifest itself after a 20- or 40-year latency period is not large enough to affect current industry conduct. When workers are still able to sue for damages, as they are, for example, in the asbestos case, the awards they receive are far larger. The prospect of enormous future court awards may be sufficient to deter current industry misbehavior.

Will the prospect of enormous court awards encourage greater efforts to prevent workplace exposure to toxic substances? The asbestos case is instructive. Dr. Irving Selikoff reported to the Labor Department in June 1981 that 8,500 to 12,000 workers who were exposed to asbestos in the past will die from asbestos-related cancer each year for the next 20 years, totaling more than 200,000 excess cancer deaths by the end of the century.[11] Asbestos manufacturers are in court to answer charges that they continued to expose workers to asbestos even after they knew of its cancer-causing potential. They are also charged with withholding information about the health effects of asbestos from the workers they were exposing.[12] More than 12,000 asbestos compensation cases, involving over 20,000 persons, have been filed against asbestos manufacturers.[13] Some have estimated the potential liability for U.S. industry at $40 to $80 billion.[14]

The courts are clearly not the proper forum for a public health crisis of this magnitude. The court system with its inevitable burden of proof on the victim is too costly, too uncertain, too lengthy, and too unfair to be relied upon as the sole avenue of recourse for victims. Moreover, the asbestos epidemic is so vast that the survival of large manufacturers and insurance companies could very well be threatened. This has provided one asbestos manufacturer, the Manville Corporation, a justification for attempting to avoid its tort liabilities by seeking to reorganize under Chapter 11 of the Bankruptcy Act.[15] If this strategy succeeds, workers suffering from asbestos-related diseases may never be compensated.

Proposed legislation would remove the problem from the courts by eliminating workers' rights to sue asbestos companies and by creating a national asbestos compensation system from which workers could recoup their losses. The system would be funded in part by the asbestos industry and in part by the insurance industry, and, according to some proposed bills, by the government. One bill would contain a presumption that workers who were employed in the industry for a certain period of time and who suffer from cancer or asbestosis would be eligible for compensation.[16] A "no-fault" compensation system thus guarantees that workers suffering from work-related illnesses would be compensated.

Workers' compensation systems, however, simply do not provide firms with a sufficient incentive to take preventive action. It is often good business to pay compensation costs rather than to pay the costs of making the workplace safe. If the bankruptcy strategy works in the asbestos case the message to industry could be that early detection of hazards and installation of control measures is not necessary. Industry may also be encouraged to avoid control measures if costly court suits are now preempted by the adoption of a national compensation system. Elimination of court suits seems especially problematic in the asbestos case, because the manufacturers appear to have been negligent in a way that should never be repeated by other firms.

The impasse is a difficult one. Its resolution requires the development of a toxic exposure prevention system that ensures compensation for victims and also contains an incentive to prevent future exposures. What seems clear, though, is that sole reliance on court suits is not an acceptable system.

Industrial Policy

A frequent criticism of regulation is that its cumulative effect on particular industries or industrial sectors can exceed the impact of each regulation considered individually. In the abstract, regulatory requirements from different agencies might have a cumulative impact less than their individual impact since compliance with one regulation might partially or totally fulfill the requirements of another regulation. Chemical manufacturers, for instance, might be required to label containers by both OSHA and the Environmental Protection Agency (EPA). By labeling according to OSHA instructions, the manufacturer could also meet the labeling requirements of EPA, and so the total impact of both regulations would be less than the sum of the impacts taken separately.

In practice, however, the requirements of various agencies might not only be duplicative, but might also be at variance. The OSHA label and the EPA label might require different information or might require labels of different sizes or colors. In these cases, an opportunity for cost saving would be lost and unnecessary costs would be imposed on industry.

A further complaint against regulation is that it can impose too large a cumulative impact on a particular industry. Each regulatory agency may pursue its own mission with single-minded zeal, perhaps in disregard of the actions of other regulatory agencies. Particular firms or industries may then be faced with a welter of regulations, which may be individually well crafted and easily affordable, but in the aggregate may amount to a considerable burden. Under the present system, regulators might not detect these problems before the fact because no single government agency is responsible for assessing the cumulative regulatory impact on particular industrial sectors such as manufacturing, or on particular industries such as automobiles and steel.

One answer to these criticisms is to incorporate regulatory objectives within the framework of a coherent industrial policy. The notion of an industrial policy is not altogether clear, but an examination of polar cases will give some idea of the range involved. At one pole, an industrial policy could simply coordinate all government actions—from tax policy to regulation—to ensure that government policy does not work at crosspurposes. One step in this direction was the publication of the Automobile Calendar under the Carter administration. This handbook compiled all the regulations and government actions that affected the automobile industry, so that regulators and other government officials would be aware of the "sunk costs" the industry already faced when considering imposing a new, incremental burden on them.

Other examples of the search for consistency and uniformity in government policy are the Interagency Regulatory Liaison Group's attempts to coordinate government policy on the regulation of cancer-causing substances and on the methodology for conducting economic impact studies. The idea of tiering regulations to meet the needs of firms of different sizes or locations was spread more rapidly through the regulatory community by this interagency group. No doubt, much further improvement could be made by coordinating regulatory agencies so that they approach new regulatory initiatives by taking into account the achievements and mistakes of other regulatory agencies. A useful way to further incorporate regulatory objectives within a narrowly defined approach to industrial policy would be for a single agency to have the authority to review and comment upon regulations as part of an effort to coordinate government actions affecting particular industries.

At the other pole, an industrial policy could be much more activist. It could, for instance, involve the creation of coordinating mechanisms, such as national or regional development banks, which could assist private financial institutions in raising capital and allocating it efficiently and equitably to all economic and regional sectors. It could allow private groups to generate government assistance for projects, such as research and development for emerging industries, that are too large or too risky for private parties to undertake by themselves, but which have the potential for significant public benefits. An activist industrial policy could also ease the transition from declining industries to emerging ones by encouraging projects that use the existing base of plant and equipment, worker skills, and public infrastructure. It could reduce the amount of capital flowing to less productive investments in corporate mergers, real estate, or silver by reducing the private costs of productive investments. It could provide mechanisms and incentives for the production of useful goods and services such as mass transit, better health delivery systems, and new forms of energy when market signals in these areas are misleading. And it could provide the context for an international trade policy which emphasized more than simply beating the Japanese in semiconductors.[17]

This activist version of industrial policy is relevant because achieving regulatory objectives requires significant investment of capital, either development capital to devise new ways in which pollution or workplace hazards can be controlled, diffusion capital to spread techniques proven useful in one industry to other industries, or direct investment in the control technology itself. Regulatory agencies do not have unlimited authority to require investment in order to attain regulatory objectives. OSHA's authority to mandate expenditures for safety and health is limited by the requirement that the technology involved be "looming on the horizon." Public funding for development or diffusion of technology to control workplace hazards could help to break through these limitations. OSHA's authority is also limited by the financial health of the affected industries. If a regulation's cost crowds out other investments that are urgently needed to prevent wholesale industry disruption, then OSHA may not promulgate it. If the industry could obtain public funds to install control technology needed for worker safety and health, this would greatly weaken the budgetary cap currently placed on OSHA's authority.

This approach provides public funds to finance regulatory objectives such as workplace safety. Like the strategy of social regulation, it has a general justification in the theory of externalities, according to which industry will not invest sufficiently in socially worthwhile activities when the costs it must bear to do this exceed the benefits it can expect to recover from this investment. Regulation has been premised on the idea of forcing industry to avoid imposing these harms by simple fiat. This command-and-control technique has been justified by the belief that industry has a moral obligation to invest enough to avoid social costs, such as those imposed by occupational injuries and illnesses, that would otherwise result from their business activity. Regulations simply specify and codify the extent of this moral obligation.

The idea that the general public should finance projects that protect worker health faces the objection that industry should not be paid to do what it has a moral obligation to do. However, the overriding objective of occupational safety and health policy should be the greater protection of workers in a manner consistent with preserving their freedom and dignity. If simply assigning a moral obligation to industry to protect worker health does not work, public financial assistance may be appropriate. In the face of the evident failure of regulation to direct more capital to the social goal of protecting workers, it is time to consider the alternative approach of encouraging industry to protect workers by making it less expensive for them to do so.

This more activist version of industrial policy presupposes that we can achieve a national consensus on priorities for social investment. To many familiar with the day to day functioning of our political system the idea that we could achieve such a consensus may appear laughably naive. For instance, a national development bank may be sound in principle, but in practice it might become simply a political fund. Groups favored by the administration in power would get the benefits while others would be excluded. Consensus would never emerge in such a context. The "public interest" would become a cover for particular private interests that currently have the upper political hand.

I will not dispute this criticism here. There is no doubt that the politics of any plan that involves targeting capital will be very complex, including major confrontations between labor and management.[18] My point is that if an activist industrial policy is adopted, one principal criterion for allocating capital should be the need for meeting social regulatory objectives, including the objective of greater workplace safety and health.

Public funding for workplace safety could be accomplished without any comprehensive industrial strategy. For instance, funds could be assigned to industries directly through the OSHA administrative budget. OSHA could then pass a regulation and allocate financial support for compliance at the same time. Alternatively, tax subsidies, accelerated depreciation, loans, loan guarantees, or direct grants from other administrative agencies could be targeted to investments in worker safety and health. All this could be done without an overall investment strategy or coordinating agency. There is some precedent for this in the provision of federal law which allows firms to finance investments in pollution control through tax-exempt state and local industrial development bonds. Approximately half (48 percent) of the pollution control expenditures required by federal law were underwritten by state and local governments in this way.

Providing public funding for investments in safety and health technology in the absence of an overall industrial policy is appealing because it avoids the need for a consensus on which industries should be favored. In addition, it does not require safety and health funds to compete on a case-by-case basis with funding for other regulatory and industrial policy purposes. It might therefore be a more direct and more politically feasible way of providing public funding for workplace safety.

This approach has a serious practical drawback, however. If industry knows that safety and health investments will be partially financed by the government, but that normal investment in plant and equipment will not, it will be tempted to disguise normal investment expenditures as safety and health expenditures. A theoretical problem of joint costs in this area makes this a likely development. Some safety and health objectives can be achieved by investments that also improve productivity. A famous example is changing the processes involving vinyl chloride, which not only protected workers from exposure to a carcinogen but also prevented loss of raw material in the course of manufacturing. Industry could easily come forward with proposals that have workplace safety as a side effect, but whose major justification is productivity. Funding safety and health investments from within an industrial policy framework would allow comprehensive evaluation of proposals having both a productivity and a safety and health justification. The price of public funding of workplace safety outside of an industrial policy is a tolerance for having a certain amount of it used for backdoor reindustrialization.

Consensus Standards

One way to improve the regulatory process is to make it easier to pass regulations that work. A major criticism of the current system has been that it takes too long to promulgate a worthwhile regulation or to eliminate or change a counterproductive one. Unfortunately, many proposals to address this real problem have concentrated, in a one-sided manner, on how to deregulate quickly. For instance, some early versions of regulatory reform legislation in the 97th Congress contained a provision for the automatic sunset of various regulations. According to this provision, regulations would no longer be in effect unless the regulatory agency went through a new rule-making process and once again established the need for them. Weaker versions of this idea would require periodic reexamination of regulations but would allow them to remain in effect during the reexamination.

Other proposals are equally asymmetrical. While most regulatory reform bills would impose cost-benefit and centralized review requirements on regulatory agencies, some versions would streamline only the deregulatory processes by exempting agency actions that repeal regulations from these requirements.[19]

A more balanced and promising suggestion for improving the process by which safety and health regulations are promulgated is to allow the affected parties greater participation in drafting the regulations. Those participating in developing a consensus on occupational safety and health regulations would include representatives of the industries and workers involved. Moving closer toward standard setting by consensus would incorporate bargaining directly into the rule-making process and would transform it from a trial-like process into a process of negotiation.

To some degree this would simply ratify current practice. Sometimes regulations are an outcome of bargaining as the regulatory agency seeks to accommodate the conflicting interests of those potentially affected by the regulation. However, the bargaining does not usually include face-to-face meetings with the affected parties under the auspices of the regulatory agency. And when a regulation is crafted to reflect a settlement that interested parties have accepted, this fact must be carefully concealed behind a facade of evidence and record building. More frequently, genuine negotiation does not occur. In these more typical cases, the trial-like atmosphere of rule-making proceedings encourages the adversaries to adopt the most extreme positions possible in hopes of forcing the regulatory agency to move in their direction. What each side can reasonably accept is not always brought out because there is no mechanism to encourage accommodation and compromise.

In contrast, regulations in European countries such as Sweden are explicitly negotiated between government agencies, business leaders, and labor representatives. Often this process of negotiation produces tough regulations while the more legalistic procedures in the United States become mired down in the construction of a record that will withstand all conceivable court challenges.[20]

This contrast between the adversary system in the United States and the consensus approach in some European countries is illustrated by their different regulations on exposure to silica in abrasive blasting operations. Breathing fine particles of sand containing silica causes a severe, progressive, often fatal lung disease called silicosis. In Great Britain the use of silica sand as an abrasive material for cleaning and polishing ("sandblasting" or abrasive blasting) has been banned since 1948. In the European Common Market Countries, the use of sand for this purpose has been banned since 1967. Sweden banned it in 1974, and its regulation cites several alternative materials that could be used to accomplish the same purpose. In the United States, despite evidence released in 1974 that sandblasters continue to suffer from silicosis and despite the presence of early available and inexpensive substitutes, sand is still used for the vast majority of abrasive blasting purposes. Preparation for a regulatory proceeding relating to a revision of the voluntary industry code covering abrasive blasting began at OSHA in 1974, and three economic impact studies have already been performed as part of this process. The revised standard itself meanwhile is no closer to promulgation than it was in 1974.

Other examples of regulatory delay are not hard to find. OSHA began to study a "right to know" standard requiring the labeling of toxic chemicals used in the workplace in 1974. The agency finally proposed one at the close of the Carter administration in January 1981, only to have the Reagan administration withdraw it in February 1981. Congressional and industrial pressure was needed to force OSHA to issue a new, and much weaker, chemical labeling proposal. The major reason for moving closer to a system of standard setting by consensus is to reduce these regulatory delays.

It should be clear that the proposal for more contact between OSHA and labor and management in working out the content of standards does not require discriminatory meetings in which one side gets the ear of a regulator while other parties are denied a chance to respond. The current administration has been criticized for repeatedly meeting with industry representatives as part of its efforts to provide regulatory relief, while excluding labor, consumer groups, and environmentalists from these meetings. For instance, C. Boyden Gray, chief counsel for Vice President Bush's Task Force on Regulatory Relief, has been quoted as saying that if a business group cannot solve a problem with a regulatory agency, it should appeal to the Task Force to get what it wants.[21] This kind of objectionable ex parte contract should be discouraged in any acceptable framework for negotiating consensus standards.

It might be profitable, however, to have all affected parties meet with agency officials, off the record, to express their views, learn what everyone can agree to, and make the necessary accommodations so that something can be done to meet a recognized need. Some regulatory reform bills in the 97th Congress moved in this direction by exempting certain agency contact with outside parties from the "sunshine" requirements of the various acts that govern rule-making proceedings. However, providing for fairness and due process in a rule-making proceeding in some ways conflicts with a negotiation framework because these procedural goals can often be satisfied only by imposing time-consuming, cumbersome, and costly safeguards. Balancing the need for new procedures for negotiation with the old procedures for fairness will be a thorny issue.

Another problem will be appeal rights for those who feel that a standard unfairly discriminates against their interests. Much of the work and delay involved in preparing a regulation under the current system results from the need to defend the standard against lawsuits routinely filed by affected parties. This appeal right should not be removed, for it is a vital democratic protection against government abuse of its powers. But it seems clear that successful negotiations leading to a consensus standard will in fact produce fewer people who feel that their interests have been so neglected that they must sue.

Collective Bargaining

The most promising local supplement to national regulation of workplace safety and health conditions is to incorporate safety and health objectives in the collective bargaining process. Bargaining could then establish local mechanisms for enforcement of federal standards. These local mechanisms are important because there are simply too few OSHA inspectors to do the job. Manufacturing and construction establishments can expect an OSHA inspection once every ten years and an average penalty per violation of $37.44.[22] This does not give management a very strong economic incentive to comply with regulations they may consider unfair or foolish.

Local enforcement mechanisms could also allow for more flexibility. Enforcement of national standards should be flexible. Rules prescribe conduct in situations that are by and large similar, or at least similar enough for the purposes at hand. But no rule can hope to accommodate itself to the endless variety of conditions present in the workplace. For example, conveyors are used in over 250 different industries, and in an almost endless variety of configurations. Detailed rules specifying safety precautions for every situation in which a conveyor might be used are simply not possible. In cases like this, performance standards that define general safety goals will be needed instead of specification standards that define precise requirements. The enforcement of these performance standards will depend heavily on flexible local enforcement mechanisms.

On the other hand, rules cannot be eliminated. They provide some degree of needed uniformity amid this diversity. For instance, exposure limits to toxic substances should be set in a uniform way to provide equal protection to all exposed workers. Moreover, sometimes diversity can be recognized in rules. In the cotton dust standard, unequal exposure limits were permitted for different processes in which workers are exposed to cotton dust, because the toxicity of the causative agent in the cotton dust varies from process to process. As a result, equal protection against byssinosis, the chronic lung disease produced by breathing cotton dust, could be provided while allowing for different exposure levels.[23] Similar flexibility can be built in by tiering regulations when hazards vary by firm size or production process used.

Although this ability to promulgate rules with built-in flexibility is limited, rigid enforcement of uniform rules in all conditions is not always prudent. For this reason, OSHA has a mechanism whereby those employers who cannot comply with a rule for technological reasons, but who can provide equivalent protection in a different way, may obtain a variance. The process is cumbersome and time consuming, and puts a large burden on the employer to show that the alternative really is as protective as the mandated procedure or equipment. But this mechanism is essential because not every possible exception to a general rule can be built into the rule itself. Moreover, if a new way of providing protection is found, there is no reason to prevent its use.

Collective bargaining can provide additional flexibility above and beyond what can be written into standards or allowed for by a centralized variance procedure. The idea is to permit unionized workers and management in particular industries or plants to develop their own compliance strategy for protection against safety and health hazards. This approach would not replace regulation, nor would it allow workers and management to collude to cut corners on safety. Rather, it would recognize the superior efficiency of the collective bargaining framework in detailing the particular methods of compliance and determining when a particular alternative allowed for in a regulation should take effect.

This approach could help to resolve one of the most difficult occupational safety and health issues likely to come up in the next few years. This is the extent to which protection against toxic substances or noise should be provided through engineering controls or through personal protective equipment. Engineering controls reduce health hazards by altering the workplace, for instance, by enclosing a process to prevent the release of toxic fumes or installing ventilation equipment to remove toxic substances from the ambient air. In contrast, personal protective equipment such as ear protectors or face masks do not alter the work environment, but place a barrier between the worker and the environment. The controversy typically swirls around technical issues such as whether equal protection can be provided by personal equipment, or whether the costs of personal equipment really are lower. A crucial additional issue may be fairness: who should bear the burden of protection? The OSHA Act clearly puts the burden on employers. This equity issue, and the greater unreliability of personal protective equipment, has convinced OSHA officials, until the current administration, to adopt a policy of requiring engineering controls to be used first, allowing personal equipment to be used only if no feasible additional engineering technique could reduce exposures to the required levels.

This general policy of engineering controls first is the most protective of workers' health and dignity. It may, however, need modification or its implementation may call for delay in particular industries or sites if its costs are extremely high. It may not be possible or desirable for this modification or delay to be initiated and imposed by a centralized office, since it would take an enormous amount of time and effort for such an agency to discover a need for an exception to the engineering controls first policy. In some cases, then, the best centralized decision might be to specify a uniform and very protective exposure level, require a general compliance strategy of engineering controls, but allow the timing and mix of compliance strategies to be worked out by labor and management when, but only when, agreement can be reached on such an alternative compliance plan within a collective bargaining framework.

This proposal faces the objection that it could expose workers to undue pressure to choose between their health and their jobs. When a compliance strategy is open to negotiation, and not simply imposed by fiat, management could present workers with a choice of going along with a dangerous and burdensome personal equipment strategy, or facing pay cuts, layoffs, or plant closings. The "consent" of workers would not be free, in other words, but would be extracted from them under threat of losing their jobs. Under a policy of always requiring engineering controls first, the option of bargaining away needed health protection is not available, and so management has to honestly face the choice of whether or not to close the plant or cut the work force. It cannot pretend hardship in the hope that workers will be frightened or powerless enough to give away a gain that is really not crucial to management, but is something that management would not mind getting. With unemployment at near depression levels, especially in the industries where the personal protective equipment strategy is likely to be attractive to management, workers may be stampeded into giving back health protection in the same way that they have been stampeded into giving back wage gains and other benefits already won under collective bargaining.

The evident risk here can be minimized by building safeguards into the process whereby worker-management agreements are reached. For instance, workers lacking strong union representation would be especially susceptible to management pressure to surrender health and safety protection. For this reason, union participation in these agreements would be essential. Moreover, OSHA would have to be part of the process. If management makes claims about the wage or job consequences of a particular enforcement plan, OSHA could require and evaluate industry documentation. OSHA would have to ensure that whatever arrangements workers and management propose do not violate the national standards, but rather are simply reasonable enforcement procedures designed to handle a particular hardship situation.

An example of flexible enforcement policy is a recent decision to allow collective bargaining to determine the mix of engineering controls and work practices to meet the OSHA arsenic standard in specific copper and lead smelters. After talks involving ASARCO, the United Steel Workers, and OSHA, an agreement was reached to install negotiated controls in an East Helena primary lead smelter. This flexible approach could save $26.9 million in capital costs if applied to all primary lead smelters and as much as $118.5 million if applied to all primary copper smelters.

Another technique to involve unions in enforcement of federal standards is the use of labor-management health and safety committees to set safety procedures and conduct routine inspections at particular work sites. Bechtel agreed to set up such a committee at a construction site at San Onofre, California, and in return California OSHA agreed to exempt the site from routine general inspections. While OSHA inspections based upon workers' complaints are not barred in this arrangement, there is an existing mechanism on the site for resolving a dispute. The threat of an OSHA inspection remains, but federal and state inspectors are freed to concentrate their routine inspections elsewhere.

While collective bargaining may have its greatest impact on enforcement of federal standards, it could also usefully supplement regulation in addressing hazards that have not yet been addressed in federal standards. This would require health and safety committees to have access to considerable information, but such access is not unprecedented. The Oil, Chemical, and Atomic Workers International Union has successfully negotiated contracts establishing health and safety committees, which are required to meet regularly to inspect or make recommendations concerning safety matters. Perhaps more important, the contract requires the company to retain at its expense qualified industrial health consultants, approved by the union, to conduct health surveys at the plant to determine if any health hazards exist. The contract guarantees the union access to company morbidity and mortality data and requires the company to pay for medical tests. All this is in addition to the company's normal industrial hygiene program.

The United Auto Workers and Steelworkers have similar contracts. Such union contracts could be expanded to include funding for additional industry-specific health research, arbitration or grievance procedures for safety disputes, a system of shop stewards to monitor safety and health conditions, and the training of these stewards and others (including arbitrators and negotiators) in health and safety hazards and control techniques.[24]

While these ideas have considerable appeal, there are several obstacles to their implementation from the perspective of both management and labor. Management traditionally has had exclusive control over the workplace. It has based this right to organize production without outside participation on the idea that property acquired in a fair way may be disposed of, within wide margins, at the sole discretion of the owner. The controversy surrounding OSHA in the last decade can be attributed in part to resentment over government and worker interference in an arena that had traditionally belonged exclusively to management. From this point of view, workers' participation in control over the workplace is objectionable in principle. Safety and health decisions are likely to be more sensitive than others in this regard because they involve decisions over capital allocation.

Unions are reluctant to bargain for safety and health because it uses up scarce bargaining capital. If management makes safety and health concessions, they will be tougher on wage and benefit issues. Union officials often express their opposition to bargaining over workplace safety by saying that safety and health are rights, and workers should not have to pay for rights by giving up wages or benefits. This argument that health and safety is a right for which no payment is necessary may technically be a fallacy: people have rights to protection against unreasonable health threats in other areas such as food and consumer products, but they are not exempt from paying for this protection through increased consumer prices. However, it is possible to argue that workers are entitled to the protection provided by federal standards and enforcement because power, mobility, information, and income are not fairly distributed between labor and management. If this equity argument is correct, then unions may be right in refusing any obligation to surrender wages or benefits in order to improve workplace safety and health.

Moreover, many union officials with firsthand experience of labor-management safety committees view them at best as ineffective and at worst as cynical ways to co-opt people. Indeed, there is some evidence that the mere existence of a joint labor-management committee makes little difference in improving occupational safety and health.[25] Why, then, should unions make bargaining concessions in order to get a joint labor-management safety committee? This would simply be to share management's responsibility for safety and health while having no real control over safety and health conditions.

In some cases, negotiated agreements on safety and health between management and labor have opened unions to liability suits from disabled workers who are blocked by compensation laws from recovering damages from management. Unions, rightly, will not participate in joint efforts to improve workplace safety unless they are exempt from these court suits.

In part, unions are suspicious of joint committees because they feel that management in the past has not taken occupational safety and health seriously enough. In their view, stricter enforcement of national standards is needed to compel management to give the problem higher priority. Only after management has been put on notice by vigorous OSHA enforcement efforts would there be an advantage for unions to bargain over safety and health.

Many union officials believe that the real problem with using the collective bargaining process to improve workplace safety and health is that unions lack power and legitimacy in the United States. The public role of unions in the United States contrasts sharply with that in countries where safety and health committees have had some success. In Sweden, for example, 90 percent of the work force is organized.[26] In the United States only approximately 20 percent of all workers are unionized, and some unions, hard pressed by the recession and plant closings, put a low priority on occupational health and safety. In these circumstances, relying more on the collective bargaining process may appear to promise little or no improvement in worker health.

There is no doubt that greater reliance on collective bargaining will increase the cost of health and safety to unions and to workers. The best response to union worries about this approach is that it is the only practical local supplement to centralized regulation that will improve safety and health conditions in the nation's workplaces. It may require labor to make concessions on wages or benefits which in an ideal world they should not have to make. But if unions do not bargain for more job safety, unionized workers will not get better safety conditions. It may be that joint committees risk co-opting workers. But without them national enforcement efforts will continue to be ineffective and unregulated hazards will not be addressed. Unorganized workers may gain little from this approach. But the best hope for improving workplace safety conditions of unorganized workers is for unions committed to bargaining for on-the-job safety to organize them.

A powerful motive to incorporate safety and health in the bargaining process is efficiency. This approach will work better than sole reliance on centralized decisions. However, greater worker involvement in safety and health provides benefits that go beyond efficiency. Worker participation programs realize a social ideal. Since health is such a fundamental interest, workers need more than the exit vote of leaving a job with intolerable risks. They need to be involved in the process by which these risks are controlled. Health and safety committees provide a way of structuring work life that would increase workers' ability to direct their lives and would reduce the feelings of anger, powerlessness, and resentment they often experience under the current system.[27]

The opposition between this social ideal and management's emphasis on property rights is stark: some people's rights to pursue private economic goals conflict with other people's rights to participate in decisions that affect their fundamental interests. Further development of a collective bargaining framework for handling health and safety, with its emphasis on accommodation of opposing interests, may produce a workable compromise despite this underlying opposition.

Conclusion

Some tentative suggestions for improving occupational safety and health policy that have emerged from the previous discussion can be summarized as follows:

1. Sole reliance on the courts and existing compensation systems is not an adequate solution. Instead, we need a toxic exposure prevention system that requires both control of future exposures and compensation for past exposures.

2. Public funding for investments to improve workplace safety and health should be incorporated in any activist industrial policy.

3. The process by which federal workplace safety and health regulations are promulgated should allow greater opportunity for negotiating and consensus building.

4. More reliance should be placed on collective bargaining and labor-management safety and health committees to enforce federal standards and to address unregulated workplace hazards.

Individually, each of the proposals seems likely to improve worker safety and health. Taken together, they could be made to reinforce each other. For instance, removing a toxic exposure compensation problem from the courts might be conditioned upon an industry's adopting an adequate correction plan to prevent future exposures. Public funding could be made available for this correction plan. Also, labor and management might be more willing to agree to a national consensus standard that contained room for the details of a compliance strategy to be worked out through collective bargaining. Finally, the risk that greater reliance on bargaining might force workers to choose between their jobs and their safety could be reduced by public funding of workplace safety investments.

More needs to be done to develop the intellectual basis for these reform proposals and to articulate the practical arrangements for implementing them. They appear to be promising enough to warrant this further work. As with all ideas for improving current institutions, the ultimate test of their validity is not their intellectual coherence, but how well they work.

Notes

The author is grateful to Mary Gibson, Karl Kronebusch, Cynthia Bascetta, and the members of the Working Group on Risk, Consent, and Air for helpful comments on an earlier draft.

1. P.L. 91–596, 84 Stat. 1590.
2. For further discussion of this contrast between individual and group approaches to risk reduction, see Mark MacCarthy, "A Review of Some Normative and Conceptual Issues in Occupational Safety and Health," *Boston College Environmental Affairs Law Review* 9 (Fall 1981–82): 773–814.
3. Bureau of Labor Statistics, *Occupational Injuries and Illnesses in the United States, By Industry, 1981* (Washington, D.C.: U.S. Government Printing Office, 1983).
4. National Safety Council, *Accident Facts*, 1980. The National Safety Council and the Bureau of Labor Statistics, the two major data sources, do not agree on their estimates of the number of industrial injuries and fatalities. The reasons for this are obscure and are only partly explained by different definitions of the work relatedness of an injury or fatality. A comprehensive discussion of these data sources and their problems can be found in K. Kronebusch, *Data on Occupational Injuries and Illnesses*, draft paper prepared for the U.S. Congress, Office of Technology Assessment project on Health and Safety Control Technology in the Workplace.
5. *President's Report on Occupational Safety and Health*, 1971, vol. 11, no. 7, p. 30. This estimate is somewhat speculative. See Kronebusch, *Data*.
6. See Peter Barth and H. Allan Hunt, *Workers' Compensation and Work-Related Illness and Diseases* (Cambridge, Mass.: MIT Press, 1980).
7. The following table displays the injury and illness rate per full-time worker:

Year	Incidence Rate
1973	11.0
1974	10.4
1975	9.1
1976	9.2
1977	9.3
1978	9.4
1979	9.5
1980	8.7
1981	8.3

In particular, the declines in injury rates in the last two years are not evidence of greater worker protection against workplace hazards. To the extent that these statistical declines reflect real changes in the workplace, they result primarily from the lowering of production rates and the laying off of inexperienced workers that takes place in recessions. For a discussion of the historical relationship between injury rates and the business cycle, see Michael Gorham, "Bum Rap for OSHA?" *Federal Reserve Bank of San Francisco Weekly Letter*, January 19, 1979, pp. 1–2.

8. National Safety Council, *Accident Facts*, 1975.

9. Economics Department, McGraw-Hill Publications Company, *9th Annual McGraw-Hill Survey Investment in Employee Safety and Health*, May 1981. The following table displays the McGraw-Hill estimates of safety investment, deflated by the GNP deflator, in billions of 1972 dollars, and as a percentage of all new investment in plant and equipment:

Year	Safety Investment	Percentage of All Investment
1972	$3.30	2.7
1973	3.41	2.6
1974	3.83	2.8
1975	3.03	2.4
1976	2.57	2.0
1977	3.08	2.2
1978	4.40	2.9
1979	2.64	1.6
1980	2.31	1.4
1981	2.63	1.6

10. Michael Baram emphasizes this criticism of the regulatory approach in "The Role of Consent in Managing Airborne Health Risks to Workers," Chapter 9, this volume.

11. Kathy Koch, "Congress Ready to Examine Asbestos Compensation Issue," *Congressional Quarterly*, February 6, 1982, p. 204.

12. An account of the evidence for these charges, which have still not been proven definitely, can be found in James Vermeulen and Daniel Berman, "Asbestos Companies Under Fire," *Business and Society Review*, no. 42 (Summer 1982): 21–25.

13. Jean Rosenblatt, "Compensating Victims of Toxic Substances," *Congressional Quarterly*, October 15, 1982, p. 769.

14. Mary Rowland, "The Asbestosis Battle," *Institutional Investor* 16 (September 1982): 345.

15. "Asbestos Firm Seeks Relief," *Congressional Quarterly*, August 28, 1982, p. 2161.

16. The bill introduced by Rep. George Miller in the 97th Congress, H.R. 5375, would provide the asbestos industry with protection against tort liability in return for participation in this compensation system. The difficulty is that this approach bails industry out without requiring or encouraging them to adopt an adequate correction plan.

17. For further discussions of industrial policies, see Ira C. Magaziner and Robert B. Reich, *Minding America's Business: The Decline and Rise of the American Economy* (New York: Harcourt Brace Jovanovich, 1982) and Barry Bluestone and Bennett Harrison, *The Deindustrialization of America* (New York: Basic Books, 1982).

18. The need for labor to be involved in the details of any industrial policy agreements is emphasized by Bluestone and Harrison in *Deindustrialization*.

19. More detailed criticisms of cost-benefit analysis and centralized review of regulatory actions can be found in Marguerite Connerton and Mark MacCarthy, "Cost-Benefit Analysis and Regulation: Expressway to Reform or Blind Alley," National Policy Paper No. 4, *National Policy Exchange*, October 1982.

20. For an account of the contrast between adversary and consensus regulation, see Steven Kelman, *Regulating America, Regulating Sweden: A Comparative Study of Occupational Safety and Health Policy* (Cambridge, Mass.: MIT Press, 1982).

21. See U.S. Congress, House of Representatives, Subcommittee on Oversight and Investigation of the Committee on Energy and Commerce, *OMB Role In Regulation*, 97th Congress, First Session, June 18, 1981, p. 53.

22. See Les Boden and David Wegman, "Increasing OSHA's Clout: Sixty Million New Inspectors," *Working Papers*, May-June 1978, pp. 43–49.

23. See the OSHA cotton-dust standard and its justifying preamble at 43 *Federal Register* 27350 (1978).

24. For further discussion of safety and health committees, see Boden and Wegman, *Increasing OSHA's Clout*, and Lawrence Bacow, *Bargaining for Job Safety and Health* (Cambridge, Mass.: MIT Press, 1980).

25. Les Boden, who is conducting an ongoing study of safety committees, reported this finding at a labor-management workshop at the 110th annual convention of the American Public Health Association, Montreal, November 17, 1982.

26. For more discussion of this contrast between safety and health policy in Sweden and the United States, see Kelman, *Regulating America*.

27. Mary Gibson relies on the idea that workers have rights to participate in workplace decisions that affect their health in *Workers' Rights* (Totowa, N.J.: Rowman and Allanheld, 1983). While I prefer to base workers' participation on a social ideal instead of a right, safety and health committees with appropriate safeguards may provide one way of realizing a right of participation.

11

The Confusion of Goals and Instruments: The Explicit Consideration of Cost in Setting National Ambient Air Quality Standards

George Eads

A number of important regulatory statutes—the Clean Air Act prominently among them—have been interpreted by the courts as precluding the explicit consideration of costs in setting health-based standards.[1] Proposals that these statutes be amended to permit (or even require) the incorporation of explicit cost considerations or the balancing of benefits against costs have been intensely controversial. Several grounds for objection have been raised, two of which will concern me in this chapter.

The first is what some see as the ethical implications of such a change. On one view, "clean" air is a "right" that "ought not" be subject to cost-benefit trade-offs. Subjecting to such trade-offs important public decisions that determine the level of protection to be provided undercuts in a fundamental way the values that the programs designed to enforce these rights are intended to achieve. On this view, it is irrelevant whether introducing trade-offs might result in higher levels of health protection: it would be wrong and should be avoided regardless of the consequences.[2]

A somewhat different but related source of opposition comes from those who hold strong pro-protection views and who believe that introducing cost considerations explicitly into basic standard-setting decisions would inevitably shift the outcome of the policy debate against these values. While this latter group sometimes uses the same vocabulary as those who hold "rights" views, their concerns are more tactical than ethical.³ Simply put, these concerns are with the consequences of the shift.

Supporters of both viewpoints draw comfort from a claimed ability to divide regulation into two distinct phases: "goal setting" and "goal implementing." In regulating airborne emissions from stationary sources the goal-setting phase encompasses the establishment of the National Ambient Air Quality Standards (NAAQS): the levels of pollutant concentration above which it is considered to be likely that some significant "sensitive population" will experience adverse health effects. The goal-implementing phase includes the establishment of the various source-specific emissions limits that determine when (and how) the NAAQS will actually be attained. In the goal-setting phase, issues of cost and technological feasibility are not to be considered. In the implementation phase, both considerations are permitted.

The attractiveness of this dichotomy to those who hold a "rights" view of airborne pollution control is easy to understand. Practicality may compel cost considerations to enter into decisions concerning how and how rapidly to implement standards, but the partitioning off of a goal-setting phase appears to permit an important element of the pollution control debate to be conducted in an atmosphere relatively free of their contaminating influence. The dichotomy's appeal to those who are concerned merely with the presumed consequences of permitting cost considerations to be explicitly considered is more complex and will be explored later.

In this chapter I advance two contentions: that the goal-setting/goal-implementing dichotomy is untenable as a matter of law, and that recent changes in our understanding of the physiological effects of air pollution effectively rule out setting the basic Clean Air Act standards without consideration of cost even if it were tenable to do so. If I am correct, then the controversy over whether the Clean Air Act ought to be amended to permit the explicit consideration of cost in the setting of the NAAQS ceases to be an argument over the ethical implications of including cost as one factor in setting these standards and instead becomes a strategic debate over the practical consequences of permitting explicit consideration of cost. Assuming this to be the case, I conclude with a discussion of the consequences of making such a change in the Clean Air Act.

The Process for Setting and Implementing the National Ambient Air Quality Standards

The Clean Air Act is an extremely complex statute. No short piece such as this one can do justice to its complexity. But for our purposes, a description of only a few of the important sections and their interrelationships will suffice.

Statutory Requirements

At the center of the act are the National Ambient Air Quality Standards. These standards, which apply to a designated set of "criteria pollutants," were originally required to be set within six months after the passage of the 1970 Clean Air Act Amendments and are supposed to be revised periodically. As of the date of this writing, only one standard—the standard for ozone—has actually been revised.[4] A proposed revision for carbon monoxide was published by the Carter administration,[5] but the Reagan administration has not moved to finalize it. The gathering of data leading to the revision of the standards for sulfates and particulates is known to be proceeding, but dates for their formal proposal are not certain.[6]

The process of setting or revising a NAAQS is supposed to begin with the collection of all scientific studies considered to bear on the relationship between ambient levels of the pollutant in question and human health.[7] Studies may involve humans or other organisms. Human studies may be clinical, that is, may involve actual experiments measuring dose and response, or epidemiological, that is, statistical studies that attempt to sort out various factors explaining variations in rates of death or the incidence of disease among large populations.

This information is to be collected, reviewed, and summarized in something known as a "criteria document." Based upon the information in the criteria document, the EPA administrator is to propose a standard which, in his or her judgment, is sufficient to protect human health "with an adequate margin of safety." The language in the report accompanying the Clean Air Act Amendments of 1970 also suggests that Congress intended that in setting NAAQS, special attention be given to sensitive populations such as asthmatics and emphysematics.[8] Courts have held that this decision, as well as the administrator's decision concerning what NAAQS finally to promulgate, is to be made without the explicit consideration of either technological feasibility or cost. That is, claims by intervenors that reaching a particular proposed standard would be "technologically infeasible," "too costly," or "too costly relative to the health benefits involved" are to be given no weight.[9]

State Implementation Plans

Once the NAAQS have been set, the primary focus of attention shifts to the states. The nation is divided into 247 air quality control regions. States are required to develop plans, known as State Implementation Plans (SIPs), showing how, for each region and for each pollutant covered by one of the NAAQS, the standards will be met. Furthermore, the statute requires that the NAAQS be attained by specific dates. Otherwise, severe penalties, including the loss of various federal funds and/or a ban on all new construction (including major renovations) in the air quality control region, will result.

In developing their plans for attainment, states are permitted to consider certain economic factors. If the cost of imposing certain control strategies would be so onerous that substantial numbers of facilities would close, with severe effects on employment, then other strategies can be chosen. But it is important to note that the sort of cost considerations permitted even at this stage of the process do not ask whether particular control strategies are "efficient," or "worthwhile." They relate solely to "affordability." Furthermore, the mere fact of "unaffordability" is not sufficient to relieve an area of the requirement to attain the NAAQS by the statutorily mandated dates.

The Link Between Setting and Attaining the NAAQS

This description of how the Clean Air Act is supposed to operate demonstrates that the distinction between goal setting and goal implementing is untenable. Once the NAAQS have been established (in theory, without regard to considerations of cost or technological feasibility), little or no leeway is open to the states under the provisions of the Clean Air Act. The District of Columbia Court of Appeals summarized the situation well in its recent decision in the case of *American Petroleum Institute* v. *Costle*:

> The goal of the Clean Air Act is to protect the public health and welfare by improving the quality of the nation's air. 42 U.S.C. Sec. 7401(b). Improved air quality is accomplished by the establishment of national ambient air quality standards (NAAQS) and by implementation thereof through state programs to control local sources of pollution. 42 U.S.C. Sec. 7410. The Act directs the Administrator to establish two types of NAAQS. Primary ambient air quality standards are "standards the attainment and maintenance of which in the judgment of the Administrator, based on such criteria and allowing an adequate margin of safety, are requisite to protect the public health." 42 U.S.C. Sec. 7409(b)(1). Secondary standards "specify a level of air quality the attainment and maintenance of which in the judgment of the Administrator, based on such criteria, is requisite to protect the public welfare from any known or anticipated adverse effects associated with the presence of such air pollutant in the ambient air." 42 U.S.C. Sec. 7409(b)(2). State control programs must provide for the attainment of primary standards "as expeditiously as practicable but . . . in no case later than three years from the date of approval of such plan. . . ." 42 U.S.C. Sec. 7410(a)(2)(A)(i). State programs that implement secondary standards must specify a "reasonable time at which such secondary standard will be attained." 42 U.S.C. Sec. 7410(a)(2)(A)(ii). *Thus, the ozone standards at issue in this case must be implemented through state plans within three years for the primary standard and within a reasonable time for the secondary standards.*[10]

The "Goals" of the Clean Air Act

The Court's use of the word "goal" in its decision in *API* v. *Costle* is instructive. For the Clean Air Act does embody goals, though not in the way envisioned by those proposing the goal-setting/goal-implementing dichotomy. The goals of the act are contained in the section referred to by the Court, the statement of congressional findings and purpose. This statement reads, in part, that the purpose of the act is:

> to protect and enhance the quality of the nation's air resources so as to promote the public health and welfare and the productive capacity of its population.

This goal statement is similar in character to statements of other goals enunciated by Congress: the elimination of poverty, the ending of racial discrimination, the achievement of full employment and stable growth. Each of the pieces of legislation setting forth programs for purposes such as these contains broad statements of ideals toward which the nation should strive and a strategy for their attainment. Some of these strategies are precisely specified; others are left deliberately vague. However, when read as part of the statute as a whole, and when their legal interconnections are made clear, it becomes apparent that the NAAQS are not goals, with the emissions control programs means of implementing them; instead both the NAAQS and the emissions control programs are instruments for achieving the broader national goal of controlling air pollution.

Clean Air Regulation in Practice: The De Facto Enshrinement of the NAAQS as "Goals"

The above account describes the Clean Air Act as it is supposed to operate. It does not describe reality. The supposedly immutable deadlines have been pushed back by Congress once already and are certain to be pushed back once again.[11] Protestations to the contrary notwithstanding, this has not been due primarily to the laziness of governmental officials or to the recalcitrance of industry, but to the growing realization that the original objectives as well as the dates for their attainment were hopelessly unrealistic. The Environmental Protection Agency, under both pro-environment and anti-environment presidents, has used whatever discretion it has — and sometimes then some — to introduce flexibility into compliance plans. Even where deadlines have not been extended, they have routinely been missed with impunity.

The menu of penalties available to EPA has not been invoked. There has, of course, been a long series of threat and counterthreat. But only one state, California, has ever had federal funds withheld.[12] Occasions have occurred (and have been allowed to pass) where EPA could, in theory, have shut down entire major industries, or levied punitive fines having roughly the same effect.[13] As Goodson has pointed out most clearly, environmental regulation in practice has been more like an extended and generally acrimonious negotiation involving EPA, the states, various industries, and Congress, with the courts often serving as the referee.[14] In this complex and politically explosive negotiating process, the NAAQS have, in fact, taken on a life of their own and, in doing so, have become something very much like goals. That is, progress in "protect[ing] and enhanc[ing] the quality of the nation's air resources so as to promote the public health and welfare and the productivity of its population" has come to be measured not by direct indicators of human health but by how many of the 247 air quality control regions have reached ambient levels that, given EPA's system of monitoring, allow them to demonstrate that they have attained the various NAAQS. The fact is that we can't directly measure how improvements in air quality, whatever they may have been, have improved human health and welfare and the productivity of our population. Science just is not up to the task. So the "instruments"—the NAAQS—become the surrogate goals.

This problem is enhanced by the highly asymmetric consequences, economic and otherwise, of being slightly below as opposed to slightly above any one of the standards. A relatively small relaxation of one of the NAAQS (as occurred in the case of ozone) might bring large sections of the country into attainment, thereby removing a cloud of uncertainty over future development and modernization plans. Conversely, minor tightening of a standard would likely throw substantial areas out of attainment, with all the attendant difficulties that would cause. All parties thus come to have an inordinately large stake in the NAAQS.

Abandoning the Notion of a Health Effects Threshold: Enter Costs

Given the weight that attaches to the NAAQS, it is little wonder that people with differing views on the value of public health protection disagree substantially over their level. It is also clear why those with strong pro-protection views would seek to keep explicit considerations of cost as far removed from the standard-setting process as possible. But is this really possible given what we know about the health effects of pollution? I believe that it is not. Experience in setting and revising the NAAQS bears me out.

If the process of attaining the NAAQS bears little resemblance to the one described in the Clean Air Act, the same can be said of the process for establishing the NAAQS themselves. Again, the reason is pragmatic. In this case, the pragmatism arises from our growing understanding of the nature of the relationship between the exposure to pollutants and possible human health effects.

The original model that underlay the NAAQS—and that permitted at least the possibility of setting these standards without considering costs—was that distinct thresholds existed below which the exposure to pollutants would cause no discernible harm, even to sensitive populations. However, as we have become better able to detect physiological responses of organisms (including human beings) to low doses of pollution and as more and more large-scale epidemiological studies have been conducted, the notion of well-defined thresholds has had to be abandoned. This abandonment has been clearest in the case of ozone[15] where, in revising the standard, the EPA administrator declared:

> The criteria document confirms that no clear threshold can be identified for health effects due to ozone. Rather, there is a continuum consisting of ozone levels at which health effects are certain, through levels at which scientist (sic) can generally agree that health effects have been clearly demonstrated, and down to levels at which the indications of health effects are less certain and harder to identify. Selecting a standard from this continuum is a judgment of prudent public health practice, and does not imply some discrete or fixed margin of safety that is appended to a known "threshold."[16]

And later:

> [A review of the scientific studies] illustrate[s] several important points: (1) all alternative standard levels reflect some risk, (2) there is no sharp break in the probability estimates that would suggest selecting one alternative standard level over another, and (3) the choice of a standard between zero and a level at which health effects are virtually certain (0.15 ppm) is necessarily subjective.[17]

How, then, was an ozone standard to be set? *No* level of ambient exposure above zero could be ruled out if consideration was given just to health effects. But an ambient standard of zero was clearly prohibitively expensive. The administrator's solution to this dilemma is worth quoting at length:

> The Clean Air Act, as the Administrator interprets it, does not permit him to take factors such as cost or attainability into account in setting the standard; it is to be a standard that will adequately protect public health. He recognizes that controlling ozone to very low levels is a task that will have significant impact on economic and social activities. *This recognition* causes him to reject as an option the setting of a zero-level standard as an expedient way of protecting public health without having to decide among uncertainties. However, it is public health, and not economic impact, that must be the compelling [not sole?] factor in the decision. Thus, the decision as to what standard protects public health with an adequate margin of safety is based on the uncertainty that *any* given level is low enough to prevent health effects, and on the relative acceptability of various degrees of uncertainty, given the seriousness of the effects.[18]

In short, costs weren't supposed to be considered, but they were—indirectly. Furthermore, this was not due to any willingness on the part of the EPA administrator to flout the law, but due to the impracticality of carrying out the law. But in order to develop a standard that would stand up in court, he was forced to pretend (though the pretense was relatively transparent in this case) that costs did not play an overt role in his decision. This he succeeded in doing, for, as we have seen, the courts upheld his decision. But in the process, the public lost the chance to examine the role that cost—as opposed to other factors—did play in influencing his judgment. All we know is that the importance of cost did not rise to the level where it offended the reviewing courts.

Is the Statutory Structure Really All That Important?

The preceding sections demonstrate that the structure of the Clean Air Act is not consistent with a goal-setting/goal-implementing dichotomy that would separate the process of setting the NAAQS from the processes designed to ensure their attainment. They also demonstrate the impossibility of excluding cost considerations, at least implicitly, from consideration in setting the NAAQS. But they also suggest that, where the realities of Clean Air Act administration clash in a serious way with the act's statutory structure, an accommodation is somehow reached. Thus, costs are considered, draconian penalties are not assessed, dates for attainment are slipped, both formally and informally—all without changing the act itself. The act thus becomes what I would term a "policy fiction," and arguments, intense though they may be, about changing the structure of the act to reflect these accommodations become arguments, at least in part, over the value of maintaining this policy fiction.

It might be contended that policy fictions are healthy—even necessary—to the smooth working of the policy-making process. By diverting conflict away from more substantive areas—like compliance methods and dates—they permit the unwieldy structure of the Clean Air Act to function. If some of the more symbolic (according to this view) features of the act were modified, attention would focus on the act's underlying structural flaws, and the whole consensus underlying the process of controlling air pollution through this structure might come unraveled.

This view has some merit. The greatest threats to the act have come when it seemed as though it would be taken literally; such as when EPA Administrator Ruckleshaus decided to refuse to grant the auto industry's request for the two-year extension in mobile source emission compliance deadlines in the mid-1970s. The very impracticality of taking the act literally has been a sort of safety valve. Amending the act to eliminate some of its more important policy fictions might remove—or at the very least tighten—its safety valves, increasing the risk of an explosion. That would precipitate a crisis that would test Congress's commitment to the current structure, a crisis that might well produce changes that the environmentalists and their allies would find extremely distasteful. However, counting on all parties to continue to wink at the law in the name of keeping the act in its current form is a somewhat risky proposition.

Other benefits of maintaining the policy fictions embodied in the act emerge from a closer look at a group of pro-protection views mentioned earlier. I said at the outset that the attraction of the fictional goal-setting/goal-implementing dichotomy to those who hold the "rights" view of airborne pollution control was obvious—at least before it became equally obvious that it is in fact impossible to exclude cost considerations from the setting of the NAAQS. I deferred commenting on why it was attractive to those who held pro-protection views based on other than a strict "rights" view. It is now appropriate to examine their motives more closely.

While the current policy fictions do not foreclose the implicit consideration of costs in setting the NAAQS, they effectively limit the weight that can be given to them. Given the inherent scientific uncertainty involved in assessing risks to human health from various levels of ambient air pollution exposure; the difficulty of deciding just which health effects, sensitive population, and margin of safety are appropriate; and the clear statutory role that the administrator's judgment is permitted to play in setting the NAAQS, there is bound to be considerable room for giving costs implicit weight in reaching the final decision. But that this weight must remain implicit provides built-in limits to how great it can be. In the case of the ozone standard revision, the area from about 0.08 ppm to about 0.20 ppm was clearly within acceptable bounds and the cost implications of choosing one end of this boundary as opposed to the other were enormous.[19] Within the middle of this range, EPA Administrator Costle stood relatively little danger of being overruled, regardless of his choice. Thus his initial proposal to relax the existing standard from 0.08 ppm to 0.10 ppm (thereby "saving" between $1.6 billion and $3.4 billion) and his final decision to relax it further to 0.12 ppm (thereby "saving" an additional $1.0 billion to $3.8 billion) were never in much jeopardy. But had he wished to go further—let us say, to 0.15 ppm or even to 0.20 ppm—the suspicion would have grown that cost was the driving factor in his decision, notwithstanding the fact that the scientific evidence showed no real break even in this range in the health risks involved. Such a move would have increased the danger of a court reversal. This knowledge, as well as his own strong pro-protection leanings, must have influenced Administrator Costle's decision.

But the view that policy fictions are desirable and even necessary to the functioning of the act and therefore should be retained has its limits. For these fictions are not costless. To be sure, EPA administrators have found a way around the absolute prohibition on consideration of costs implied by the act (at least as the courts have interpreted it), and the courts, while continuing to cite precedents that would appear to compel overruling these administrators' decisions, have nevertheless upheld them as "supported by a rational basis on the record" and held that the process by which these decisions were reached, "[a]lthough not a model of regulatory action," was not so flawed as to warrant invalidation of the final standard.[20] But to arrive at this outcome, decision-making processes have had to be artfully concealed and evidence that has figured prominently in decisions has had to be disowned. It has been impossible, for example, to adopt a decision-making procedure such as that proposed by Richmond, in which the evidence upon which a NAAQS will be based, scientific and otherwise, is formally arrayed, evaluated, and presented to the administrator in a form amenable to the clear weighing of trade-offs.[21] It also has been made impossible to demand of the administrator a rational, documented accounting of the weight he or she has given various factors in reaching the announced decision. We see only the shadow, not the substance.

To sum up the debate: the current structure of the act, which makes difficult, but does not effectively prohibit, cost considerations in setting the NAAQS, clearly works to reinforce pro-protection values, at least as those values are reflected in the NAAQS. To the extent that NAAQS are valid surrogates for the primary goal of the Clean Air Act, which is, to repeat, "to protect and enhance the quality of the nation's air resources so as to promote the public health and welfare and the productive capacity of its population," then the public interest might be said to be served. But if the NAAQS are faulty surrogates, then this outcome may not advance the public interest.

But this returns us to the enshrinement of the NAAQS as de facto goals, not the instruments they originally were designed to be. In my view, the value of focusing so much attention on the NAAQS needs to be a subject of serious national debate. But the current policy fictions embodied in the Clean Air Act make such a debate all but impossible, since they equate such a questioning with a lack of concern for clean air and all the obvious benefits it generates. This, perhaps, is the greatest cost of maintaining these policy fictions.

Notes

1. See, for example, *Lead Industries Assn.* v. *EPA*, No. 78-2201 [14ERC 2089] (D.C. Cir. June 27, 1980).

2. The clearest and most complete statement of this "rights" view and of its implications can be found in Steven Kelman, *What Price Incentives?* (Boston: Auburn House, 1981), Chap. 2.

3. This view seems to lie at the heart of the objections raised by the National Commission on Air Quality in its report, *To Breathe Clean Air* (Washington, D.C.: Government Printing Office, 1981), Part 3, Chap. 1. See also the exchange between the author and Senator Hart,

the Chairman of the Commission, in *Clean Air Act Oversight: Hearings Before the Committee on Environment and Public Works, United States Senate, 97th Cong., 1st Sess., June 5, 9, 11, and 22, 1981, Part 3*, pp. 213–17.

4. The final rule revising the ozone standard was published in the *Federal Register* on February 8, 1979.

5. *Federal Register*, August 18, 1980. The administration's unified Agenda of Federal Regulations, published in October 1982, shows the carbon monoxide standard being made final in October 1982, but this (self-imposed) deadline was not met.

6. The Agenda of Federal Regulations shows the revised sulfur oxide, particulate, and nitrogen oxide standards being proposed in mid- to late 1983.

7. A more complete description of this process may be found in *To Breathe Clean Air*, Part 3, Chap. 1.

8. On the issue of "sensitive populations," see Robert D. Friedman, *Sensitive Populations and Environmental Standards* (Washington, D.C.: The Conservation Foundation, 1981).

9. See, for example, *American Petroleum Institute v. Costle*, 16 ERC 1435 (1982) at 1441 and 1442.

10. *American Petroleum Institute v. Costle*, 16 ERC 1435 (1980) at 1438 (emphasis added).

11. The second deadline for nationwide attainment of the NAAQS was December 31, 1982. It is no secret that this deadline was missed. According to EPA, as of December 1982 only one state, North Dakota, is fully in compliance. Some 600 counties, one-fifth of the nation's total number, are out of compliance for at least one pollutant. Cited in "Clean Air Force," *Wall Street Journal*, December 15, 1982, p. 30.

12. These were restored by President Reagan just prior to the 1982 elections. *New York Times*, October 13, 1982, p. A22.

13. We came close once, however. EPA Administrator Ruckleshaus declared in 1973 that the automobile industry had not met the statutory requirements for a two-year extension of the mobile source emissions controls and denied its requests. Ruckleshaus was sued and ordered by the Supreme Court to reconsider his decision, taking "broader" factors (arguably, factors not contemplated by the Clean Air Act) into account. He did so and granted the extensions, though with conditions attached. See *International Harvester v. Ruckelshaus*.

14. R. Eugene Goodson, *Federal Regulation of Motor Vehicles: A Summary and Analysis* (Lafayette, Ind.: Institute for Interdisciplinary Engineering Studies, Purdue University, March 1977).

15. While the undermining of the concept of a well-defined health effects threshold may be clearest in the case of ozone, the situation is by no means much better (if it is better at all) in the case of the other pollutants for which NAAQS have been established. Indeed, in its proposal to revise the ozone standard, EPA quoted the following passage from a 1974 National Academy of Sciences study concerning the general viability of the threshold concept:

> in no case is there evidence that the threshold levels have clear physiological meaning, in the sense that there are genuine adverse health effects at above some level of pollution, but no effects at all below that level. On the contrary, evidence indicates that the amount of health damage varies with the upward and downward variations in the concentration of the pollutant, with no sharp lower limit. (EPA, Proposed Revisions to the NAAQS for Photochemical Oxidants, *Federal Register*, June 22, 1978, p. 26965.)

16. Ibid.

17. Ibid., p. 26967.

18. Environmental Protection Agency, Revisions to the National Ambient Air Quality Standard for Photochemical Oxidants, *Federal Register*, February 8, 1979, p. 8213 (emphasis added).

19. According to White, the difference in annual costs of meeting the most stringent (0.08 ppm) as compared to the least stringent (0.20 ppm) of these standards ranged from $3.1 billion to $13.1 billion, depending on who made the estimate (EPA or the interagency Regulatory Analysis Review Group) and the model used. Lawrence J. White, *Reforming Regulation: Processes and Problems* (Englewood Cliffs, N.J.: Prentice-Hall, 1981), p. 62.

20. *API v. Costle*, 16 ERC 1448.

21. Harvey M. Richmond, "A Proposed Risk Analysis Framework for National Ambient Air Quality Standard," 1981.

12

Achieving Air Pollution Goals in Three Different Settings

Clifford S. Russell

Public and academic debates touching on risk and consent in air quality management are usually dominated by questions about what we are or should be trying to achieve: what outdoor air quality standards should we be aiming at? What levels of hazard are acceptable in the workplace? How much side-stream smoke can we tolerate in offices and public places? These questions dominate in part because they involve human health, an intensely emotional matter. Beyond that, such questions bring us face to face with another deeper and even more intriguing level of question, the level on which we ask ourselves how to decide the how-much questions. For example, is cost-benefit analysis ethically defensible and politically meaningful? If we do want to quantify benefits in money terms, how can we best do so? How (or should) we take into account the apparent difference between involuntary exposure to ambient air pollution and the largely voluntary choices of individuals about where they work and what they work at? Finally, answers to questions about the goals of public policy for air quality management seem to drive the system in a rather obvious way; if we want a higher goal we have to pay a higher cost. Therefore, it is usually assumed that deciding on the goal must leave only a few technical details to be wrapped up before we can get on with the business of achievement.

This view, while common, is wrong and dangerous. It is wrong because deciding on how to achieve a chosen public policy goal may be just as hard as deciding on the goal itself. It is dangerous because by ignoring the technical and political difficulties of choosing what may be called, briefly if infelicitously, implementation systems, we may enormously increase the cost of achieving the goal, or even make the goal unachievable.

The view taken in this chapter is that choice of goal and choice of implementation system are separable, even though equally difficult, problems. The discussion here will take as given a goal or standard for ambient environmental quality. A conceptually preferable, if operationally quite hopeless, view is that goal and implementation system must be chosen together, the results of a grand meta-benefit/cost analysis.

The fundamental problem in passing from goal choice to goal achievement in air pollution and many other public policy areas is that achievement requires actions by many individual actors (households, firms, and other levels of government), actions that will almost always be against the narrow self-interest of the actors. Deciding how to achieve an air pollution goal, then, requires three closely related, subsidiary decisions:

- what actions it is necessary for the individual actors to take;
- how to induce the actors to take these actions;
- how to find out if the actions are being taken.

Each of these decisions is, in turn, difficult, and three different sorts of difficulty are involved:

- technical matters, such as identifying which sets of possible actions could physically result in achievement of the goal, but more generally, the technical problem of obtaining, manipulating, and interpreting large amounts of information;
- the confounding effects of adjustments by the actors to the initial choices, such as the decision to delay replacing an old plant or to fund research on new pollution control technology, in response to the incentives contained in one or another implementation system;
- ethical and political judgments about the implementation system itself, quite independent of its ability (or lack thereof) to achieve the goal.

Let us consider further the nature of these difficulties as background for later discussion of some of the major contenders for pride of place among implementation systems. This preliminary discussion will be confined to the outdoor air pollution case, where the points are most straightforwardly illustrated. The first difficulty we find is identifying the set of individual actions that are both feasible for the individual actors and capable of producing the desired result. In the outdoor air pollution context this means finding the set of feasible discharge combinations; that is, the combinations of discharge levels at all the sources in the region (or regions) of interest that can be achieved by the dischargers and that do, taken together, produce the desired ambient quality goal. Notice that making this identification requires us to know something both about the capabilities of dischargers and about the characteristics of the regional atmosphere and topography, the latter items being necessary to predict how a given pattern of discharges from the region's sources will be transported, transformed, and diluted after leaving the source stacks and what the resulting ambient quality will be at particular monitoring points in the region.[1] For purposes of ex ante policy analysis, the necessary information must be formalized in a more or less complicated and sophisticated mathematical model of the region. The simplest such model assumes that a given percentage reduction in discharges of a pollutant from all sources produces the same percentage reduction in ambient concentrations of the pollutant at all monitoring points. But this model presumes a particular, and usually particularly costly, set of orders to sources: equal percentage reductions.[2] Finding the full feasible set, even in an approximate form, requires a much more sophisticated model of the regional atmosphere.[3]

A second technical difficulty is choosing preferred members from the feasible set of combinations of source control orders given an accepted standard of judgment.[4] Here, the most common standard, even among noneconomists, is total cost to society. Finding the least cost method of achieving the desired ambient quality goal requires all the information necessary to identify the feasible set plus data on the cost to every discharger of reducing discharges over the technically possible range for that source. (This information for a single discharger will be referred to as the cost-of-reduction function.) The lowest cost combination of orders is referred to by economists as the efficient policy.[5]

An additional technical problem opens up when we recognize that even if two different feasible combinations of discharge limitation orders imply the same total costs, the costs borne by particular firms, industries, subregional areas, or other identifiable parts of the population would in general be different for each. Predicting the distribution of costs as opposed to the total costs is difficult enough if we confine our interest to initial incidence, that is, to the firm, group, or government unit that first pays the cost. It becomes for practical purposes impossible if we want to identify ultimate incidence. To do this we would have to have complete information on the structure of demands for goods and services, the conditions under which those goods and services are supplied, how the tax system works, and how all this fits together simultaneously in the market economy.[6]

The second class of difficulty lying in the way of a useful appraisal of alternative methods of passing from collective goals to their achievement is the confounding effect of individual reactions to any particular implementation method. That is, we cannot judge an appraisal sufficient if it looks only at the operation of the method in the world as it is. Different methods will present different opportunities for profitable adjustment to self-interested firms, households, and even to units of government.[7]

It may be possible to say something about the desirability of some of those adjustments. For example, some methods may provide continuing incentive for pollution dischargers to search for new production and waste treatment methods that will result either in lower cost achievement of the chosen ambient quality goals or in higher ambient quality at the same total social costs. Other policies may discourage this search; and still others may be neutral in this regard. While it is anything but a straightforward matter to identify, measure, and compare the total costs and benefits (including the costs of research and development as well as of application of technology itself, and the benefits of the resulting quality levels) of such alternative future paths, many would agree that all other things equal it is better to encourage than to discourage the search for environmentally cleaner technologies.

Other types of adjustments are possible, of course, and a complete catalog would be impossible a priori, but a few general rules will give the flavor. If an implementation method differentiates among sources on the basis of location, it will produce an incentive to move from high cost to low cost places, though this will be damped to the extent that high cost and high quality (and low cost and low quality) areas tend to coincide. A method, such as the one we have in the United States today, that puts more of the burden of achieving the goal on new sources and facilities will tend to encourage longer lives for old sources and to discourage the appearance of new facilities. A method that changes the relative costs of final products by increasing the costs of some industries proportionally more than the costs of others will tend to discourage consumption of the products of the former. Even if consumers' tastes do not change, we would expect to see changing life-styles as a result of a pollution control policy. And we would expect that production levels, and hence discharges, of the higher cost industries would tend to be lower than would have been predicted in the absence of the policy.

These last observations suggest that there is another side to the adjustment coin. As adjustments take place, and as other, unrelated outside changes impinge on the economy, it is very likely that at least the efficient way of achieving an unchanging ambient standard will change. Indeed, in general even the feasible set of discharge limitation orders will change. A set of regulations that produced the desired goal at least cost in year one might very well not produce the goal at all in year 20, as growth, shifting locations of production and residence, changing technology, and changing product mix change the spatial pattern of pollution discharges.[8] Thus, implementation is a continuing, not a once and for all, problem.

The third class of difficulty we face in choosing among implementation methods involves ethical and political suitability. A system that offends commonly held ethical norms will not be used no matter how attractive its efficiency, distributional, and dynamic incentive properties. But a system might also be rejected because its distributional characteristics were considered politically unacceptable even though it was not ethically objectionable. The problem here is twofold: first, to identify the ethically and politically relevant features of alternative systems; second, to make useful analytic statements about the acceptability of the systems.

For example, at the simplest level, it is clear that the matter of cost distribution is intimately linked to the political viability of alternative ways of meeting collective environmental goals. Because we choose a distribution of benefits when we choose the goals, and because we have no mechanisms (other than the very creakiest ones) for redistributing incomes, a choice of cost distribution implies a fixed pattern, in the short run, of net benefits for that broad area of environmental policy.[9] If an analysis of the distribution of costs and benefits shows that a majority of voters or the members of some powerful or vocal voting block will probably incur net costs from the policy, one would certainly be tempted to predict a rocky road for it.[10] Thus, it is not necessary to judge the distribution on the equitable/inequitable axis in order to say something useful.

The ethically relevant features of the alternatives are important in themselves and also help to determine political acceptability. It is, however, extremely difficult to provide useful analysis of these features. First, there appears to be no agreement on a list of ethically relevant features. Second, for any given feature it seems often to be the case that the direction of ethically "better" is itself in dispute. For example, one segment of society feels strongly that an ethically acceptable pollution control policy must stigmatize polluting activities.[11] This view is sometimes based on religious views about nature, sometimes on Lockean views about causing harm to others (see Chapter 5), and sometimes on opposition to industrial or market activity of any kind. Others disagree that polluting, as opposed to disobeying laws about polluting, is a bad to be stigmatized. These people point to physical laws of conservation of mass and energy to justify the view that some waste, hence some pollution, is inevitable in any production or consumption process.[12] To make an ethical issue of pollution is for them as unhelpful as to make an ethical issue of the human need to eliminate body wastes.

Or consider the matter of choice. For the sake of illustration, we might agree that one ethically relevant issue is the extent of choice allowed to pollution dischargers in responding to the orders issued by government to achieve the chosen goal. Agreement would not, however, exist on the question, is more choice better or worse? Roughly speaking, those who would stigmatize pollution would narrow choice. Those who see no ethical issue in the fact of pollution would expand choice, generally arguing that to do so lowers the total cost of achieving the standard.[13]

For the most part the central notions around which this volume is organized—risk and consent—are less clearly applicable to choice of an implementation system than they are to choice of the ambient quality goal itself. However, risk is an attribute of the implementation system because no system can guarantee the goal, and some leave more room for failure than others. Thus, if polluters have substantial choice they may choose to act in such a way as to frustrate the predictions on which the implementation system is based.[14] The prospect of such actions poses a risk of failure to reach the quality goal.

Because implementation systems operate on the sources of pollution, the issue of consent can enter the discussion only if we are willing to stand it on its head relative to its orientation in the other chapters. Thus, we might ask which implementation system involves the greatest or the clearest amount of consent on the part of those sources. In such a context choice and consent seem to amount to the same thing, and the comments above on choice as an ethical value could be repeated here, though the use of the word "consent" might stimulate the disputants to raise their voices a few decibels.

The above discussion was intended to convey the difficulty of choosing how best to go about achieving accepted public policy goals. The rest of this chapter will be organized around a more careful analysis of five dimensions along which alternative implementation systems may be considered, and to some extent judged. These dimensions will be:

1. Efficiency. The efficient implementation system achieves the chosen goal at least cost. This dimension is almost always interpreted in a static sense and that will be the approach here. "Static" means, as a practical matter, that we assume an unchanging goal and allow only for the first round of reaction to the implementation orders or incentives.
2. Information intensity. This criterion involves an attempt to measure, at least qualitatively, how much data and what level of predictive modeling skills must be available to the pollution control agency to use the implementation system in question.
3. Ease of monitoring and enforcement. This refers to the relative difficulty of making and interpreting the measurements necessary to judge compliance, prepare bills, or audit self-reporting. The chapter also deals more cursorily with the question of penalties and penalty assessment.
4. Dynamic characteristics. This dimension involves, first, whether the instrument provides continuing incentives to polluters to develop and put in place environment-saving (discharge-reducing or -transforming) technology. Second, it involves how flexible the im-

plementation system is in the face of change—both that stimulated by the system itself and that occurring independently.
5. Political and ethical appeal. This is not a single dimension at all, but rather an umbrella covering a number of more or less analytically tractable notions. One is the distribution of costs; a second is the apparent risk of failure (or the apparent scope for perverse choice); a third is the degree to which polluting behavior is stigmatized; and a fourth is the extent to which choice is allowed to sources of pollution.

The alternative implementation systems to be considered will include the most important representatives of the two major types: those that specify results dischargers are required to achieve (often referred to as command and control) and those that offer the discharger an economic inducement to act in some way. The results specified might be the installation of particular equipment, the use of particular inputs, the reduction of prepolicy discharges by certain amounts or percentages, or the maintenance of discharge levels below some allowed maximum. These are the major alternatives usually discussed, but the list does not exhaust the set of possibilities. Each possibility will show up differently when measured along our five dimensions. Further, it should be said at once that the specification of results is often backed up by economic penalties (fines) for noncompliance. Economic incentive systems require either the payment of a bill or the observance of permit terms (even though the permit may be salable property). Thus, neither system is found in a pure form, but it is convenient for discussion to draw the line between them in such a way that our attention is concentrated on the principal interactions between regulatory agency and dischargers (and among the dischargers themselves where permit trading plays a role).

Before getting on with this discussion, however, one further source of complexity must be stressed. Our illustrations have so far been drawn from the problem of managing outside air quality. This is the policy context in which the relative merits of economic incentives and command and control regulation are most frequently debated. When the policy context changes to workplace air quality or to side-stream smoke, certain important features change as well. Even if the policy goals are defined in the same terms in each context—as maintenance of some air quality standard at monitoring points located through the "environment," be it metropolitan area, steel mill, or restaurant—those changes in setting imply changes in what can be said about the usefulness of particular implementation systems.

The major differences among the contexts concern: the number of contributors to the air quality problem; the identity of the contributors; and the presence or absence of a population of potential monitors, individuals with the ability, motivation, and opportunity to check up on the maintenance of the standard.[15]

These differences are summarized in Table 12.1.

After this rather extensive introduction it is now possible to turn to more specific commentary on implementation systems and how they compare along the five major dimensions of judgment.

Efficiency and Information Intensity

These two dimensions are necessarily considered together. As a generally applicable rule, significantly more information must be available (usually, but not always, to the regulatory agency) if a policy result is to be achieved efficiently than if it is to be achieved in any of the infinite number of nonefficient but feasible ways. Another way of putting this rule is that to save money in the actual implementation actions we must spend money to discover the cheapest set of actions to order or encourage. This unfortunate situation holds whenever the actions of more than one actor influence the measurement that must meet the collectively chosen standard, and when the influences of different actors on the measured quantity are different. The rule applies equally, though with some qualifications, to command and control orders and to economic incentive systems; and, indeed, its major force is to break down a distinction between those alternatives that has often been made in the environmental economics literature.

From the summary of differences in context given in Table 12.1, it seems that this rule is most important for the outdoor air quality problem. The combination of many sources of pollution and differential effects of each source on the ambient results at the monitoring points implies that for implementation by either orders or incentives to be efficient the orders or prices must be individually tailored to each source.[16] Thus, in general it will be inefficient to announce rules requiring equal reductions of discharges (either in amount or percentage terms) or equal postpolicy discharges at all sources. Similarly, it will be inefficient to impose an equal emission charge (price per unit of pollutant discharged) at each source.[17] This means, as a practical matter, that the efficient implementation system in the outdoor air pollution case cannot be found by trial and error but must be identified a priori through the use of mathematical models.[18] These will have to reflect costs of reduction at the sources and the differential effect of the regional atmosphere and topography on the emissions from each source once outside its stack. Hence the large initial information costs of seeking efficient systems.

Marketable permits, usually included as examples of an economic incentive system for environmental control, also reflect the connection be-

Table 12.1 Differences in Context: Three Air Quality Problems

Problem Context	Usual Number of Polluters	Identity of Polluters	Availability of Potential Monitors	Do Different Polluters Affect Measured Ambient Pollution Differently?
Outside Air	Many	Firms, Households, Government Units	None	Yes
Workplace Air	One	Firms (usually)	Work force	Only one polluter
Side-stream Smoke	Many	Individuals	Work force (offices) Customers (restaurants) None (Lobbies)	Can assume no as first approximation

tween efficiency and information intensity, but do so in a slightly different way. Two types of marketable permits are possible in the outdoor air pollution case: permits to discharge a quantity of pollution (possibly further specified as to location and timing of discharge); and permits to increase the ambient pollution level at particular monitoring points. Under a discharge permit system a discharger would have to hold (own) permits covering the quantity of pollutant he was discharging, probably defined as weight per unit time. Under an ambient permit system a discharger would have to hold a portfolio of permits covering the ambient quality deterioration his discharges caused at each monitoring point in the region, probably defined in concentration units over some averaging time.[19]

The simplest marketable permit scheme involves permits to discharge pollutants, which are freely tradable among sources across the entire region. The resulting market would be simple, in the sense that it would involve a single price for, say, a unit of SO_2 discharge per unit time. Each discharger could buy or sell a unit of discharge per unit time at that price regardless of the size of his existing discharge and of his location. The information requirement for such a system arises from the need to ensure that the ambient standards are met. In order to guarantee this at minimum information cost, the initial permit total for the region must be small enough that no pattern of trades could result in a concentration of discharges that violated the ambient standard. This permit total will generally be small relative to alternative solutions that rely on some advance knowledge of where the discharges will occur. That is, a system structured in this very simple way would involve large discharge reduction costs.[20]

These costs can be avoided only by paying higher information costs. For example, with the same kind of regional model described above, including cost-of-discharge-reduction functions, the equilibrium in the regional rights market could be predicted. (This would be the same as the situation that would exist for a uniform emission charge.[21]) The regional total of permits issued could then be set equal to the total under this predicted equilibrium.

Other refinements to a discharge permit system can be designed. These might involve subregionalization or "trading ratios" (which make a permitted pound of, say, SO_2 discharge from point A "worth" only some fraction of a pound at B or more than a pound at C). But these expedients do not vitiate the fundamental point: that efficiency is information intensive. Subregional restrictions on allowable trades, for example, constrain the system away from the least cost solution, while the establishment of trading ratios involves both centralized information (for calculation from dispersion models of ratios applicable to all possible trades) and decentralized complications for the traders themselves.[22]

The efficient solution to the regional problem can in principle be attained via a system of tradable ambient pollution rights.[23] The agency can issue, by one or another means, permits for each monitoring point up to the standard. (That is, if the standard is 75 micrograms per cubic meter on average over 24 hours, and permits come in 0.01 microgram units, 7,500 permits could be issued at each monitoring point.) The number of separate markets (the number of separate goods to be traded and hence the number of prices) equals the number of monitoring stations at which the permits are defined. Because an action by a particular discharger to increase or decrease his discharge will in general have an ambient impact at every monitoring point, decisions to buy and sell permits are not simple matters of balancing cost vs. gain in one market. Rather, the cost of discharge reduction (gain from discharge increase) must be balanced by the source against the sum of gains from permit sales (purchase costs) across all the markets, the permit quantities to be sold (bought) depending on the dispersion characteristics of the regional atmosphere. Because all dischargers face the same kind of calculation when contemplating any adjustment, it is possible that such a market will not prove practical, or at least that some sort of centralized intervention will be necessary to facilitate adjustments.

Thus, in the context of outdoor air pollution, when the link between static economic efficiency and information intensity is taken into account, there is very little to choose between orders and incentives as instruments for ambient quality achievement.

In the workplace context, the situation is entirely different. There is only one source in the decision-making sense, however many sources (vats, furnaces, stacks, machines) there may be physically. This implies that the problem of regulation is much simpler in principle than for outdoor air quality. The regulator can simply announce to each industry or plant the required ambient standard and then arrange to monitor its observance. This simplification further implies that a simple announcement of the standard can bring about the least cost solution, since each firm can be presumed to minimize its cost of compliance. Notice further that this efficiency is bought at small cost in information once the standard is chosen. There is no need to know either cost functions or dispersion characteristics.

If, however, the agency insists on specifying particular technologies for production or indoor air pollution control it can manage to avoid the efficient solution and to multiply its information headaches manyfold. Why would it want to do this? One possible answer relates to monitoring and is considered in the next section.

Further, it is hard to see how economic incentive systems, at least of the usual sort, have any potential here. Marketable permits, however defined, have no meaning, unless one is willing to contemplate markets involving each firm and its workers (or their union), for these are the only interested parties. Finding a charge, whether on ambient quality or offending discharges, that achieves the desired result could be an enormous task, and there would seem to be no benefit in the static view from undertaking it.[24]

More adventurous alternatives might be imagined, however. One candidate is some version of a liability rule. That is, no standard would be promulgated, but firms would somehow be held accountable for their employees' health.[25] Something like this currently operates, albeit apparently unsatisfactorily, through the experience rating system of large employers under state workmen's compensation laws. Two difficulties suggest themselves. One is that liability schemes, in operation, seem to be rather clumsy and blunt instruments, and especially so when lagged health effects are involved. A second is that such approaches do not help with day-to-day discomfort that never amounts to or causes morbidity or mortality. It may be that more subtle and sophisticated liability or insurance schemes can be (are even now being) devised. I think we have a right to be skeptical, however.

At first glance, the problem of side-stream smoke in public places appears very similar to the outdoor pollution problem under "mixing-bowl" atmospheric conditions. In any enclosed public space there are (potentially) many sources, and with poor ventilation each source may add roughly equally to the uniform ambient pollution level. Such a characterization suggests the possibility that a uniform "emission" charge (fee per cigarette) might be competitive on the efficiency/information dimension with regulatory orders. Further examination, however, reveals this analysis to be quite misleading—even without probing the assumption of poor ventilation.

Accepting for the moment that an ambient standard, defined at specific monitoring points, may reasonably be taken as the policy goal, the significant difficulty with the smoking problem is that the cast of "dischargers" is constantly changing.[26] A given fee per cigarette smoked might produce the desired standard at one moment and fail to produce it the next, as random lighting up occurs and as people come and go.[27]

But even orders of the type most obviously analogous to those useful in the outdoor air pollution case do not seem very promising in this problem context. For example, a limit on cigarettes per day or per hour per person in the public space might on average meet the standard. But temporal "hot spots" could be expected (the analogues of spatial hot spots in the discharge permit case discussed above). And a rule limiting the number of cigarettes going at any one time leaves open the decision of whose cigarettes those will be. This decision would be a very difficult one for a group to solve on its own, especially groups only temporarily thrown together.

We can also probe the assumption that ventilation is so poor that uniform mixing occurs. In fact this may only infrequently be true, and most smokers will create spatial as well as temporal hot spots.

Taken together, these considerations suggest that the side-stream smoke problem requires quite a different approach. The major alternatives seem to be: a flat prohibition on smoking in particular types of places, emission zoning (or segregation) such as no-smoking sections of aircraft, and possibly ventilation requirements stringent enough to reduce the "hot" area around each smoker to a very few square feet.[28] The first two alternatives are common and hard to fault. But the concept of efficiency is of very little use as a dimension of judgment in the circumstances.

Ease of Monitoring and Enforcement

These are difficult areas of inquiry, and to do them even modest justice would require at least another paper of this length. On the other hand, standard discussions tend to rest on naive and misleading assertions, and it should be possible in a reasonable space to introduce some important clarifications. To this end, let us first consider some of the faulty arguments in the outdoor pollution case and then see what is left by way of positive contribution.

An extreme simplification of the monitoring and enforcement problem is the voluntary compliance view. Here the idea is that firms (and municipal agencies operating incinerators and sewage treatment plants) are by and large good citizens, and that once the regulatory requirements are made clear, they will comply. Monitoring is not a widespread necessity, for the few malefactors will be so obvious that they will practically convict themselves. Then recalcitrance will be appropriately punished by stiff civil or even criminal penalties.

Malefactors there have certainly been, as anyone who even casually follows environmental news can testify. More difficult is the question of whether, with the exception of these relatively few but widely publicized cases, day-to-day compliance has generally been good. Evidence on this point is lacking, but such fragments as do exist suggest that a significant fraction of all point sources are out of compliance for substantial periods of time every year.[29] Thus, one cannot prove a case either way. But certainly one cannot reject out of hand the argument from discharger self-interest: that noncompliance will be in a source's interest so long as the expected savings from violation exceed the expected penalties from detection and punishment.

A second line of policy development attempts to come to grips with this argument. Here, compliance with regulations (technology-based discharge limitations) is to be encouraged by a noncompliance penalty. This idea, which originated as a practical policy in Connecticut and came to Washington with Douglas Costle, former administrator of EPA, is now part of the Clean Air Act.[30] It allows EPA administratively to assess, on a source not complying with discharge regulations, a penalty equal to what the agency calculates the source would save through its noncompliance. But the difficulties of defining and measuring noncompliance in a world of production level variations, production and treatment process upsets, imprecise, unreliable, and noncontinuous monitoring equipment, not to mention legal questions about access for monitoring purposes, are given very little, if any, attention in this line of policy recommendation.

Subtler analysts have addressed the monitoring problem by invoking the IRS analogy. Violations will be reported by the dischargers themselves, who will be required to monitor their own emissions, to report regularly to the responsible agency, and to identify periods in which they have been operating in violation of standards.[31] The incentive not to misreport the self-monitored data is supposed to come from a system of (possibly random) performance audits conducted by the responsible agency or its nominee.[32] Evidence of misreporting would lead to civil or criminal penalties.

The difficulty with self-monitoring is that the IRS analogy is a poor one. The IRS has two advantages in administering our system of self-reported income taxation that an environmental agency will inevitably and unavoidably lack:

- The IRS has access to an extensive paper record covering most if not all income for most individuals and firms. Its audits consist largely of checking that record against self-reported income and deductions. Where the IRS's own records (W-2s, 1099s, etc.) are inadequate, it can require proof from the taxpayer that claimed transactions really took place, and it can check bank account transactions for evidence of unreported income.
- For the largest part of taxable income, it is to the advantage of taxpaying firms or individuals to make sure that the IRS has a complete record for other firms and individuals, because one taxpayer's income is another taxpayer's deduction (cost).

But, in the absence of continuous agency monitoring, there is no independent record of what self-monitoring dischargers are emitting. Once discharges have gone up the stack or out the pipe they have vanished from an enforcement point of view leaving no record in the world.[33] Further, it is not in general in the interest of any particular firm or individual routinely to provide independent evidence of discharges by another source.[34]

We have now rejected three purported solutions to the two basic problems of monitoring and enforcement in the outside air pollution context: finding out what the actors are doing and punishing those that are doing wrong. Let us go back and ask why these are difficult problems and see what, if anything, can be said about their relation to the design of implementation systems.

The features of the monitoring problem that combine to make it especially challenging seem to me to be five in number. First, as already noted, all emissions are fugitive in the sense that once outside the source's stack they are lost to measurement. They leave no trail unless some human agency intervenes. Thus, we cannot monitor at our leisure if we really want to know what is and has been going on. Second, discharges vary randomly because there are random equipment breakdowns, shifts in product mix or input quality, and changes in production levels at the sources. These variations, it must be stressed, need not reflect any intention the discharger might have to cheat; even the best corporate citizens can suffer a breakdown of a precipitator that results in vastly increased emissions. This randomness has itself two implications:

1. We cannot usefully think of discharge limitation orders as simple fixed numbers. The appropriate orders for a region must take into account source variations and the probability of ambient standard violations. In addition, the orders must recognize in one way or another that the source in adjusting to the order (or to an economic incentive) must balance probability of violation against cost of controlling or narrowing its range of variation.[35]
2. The rules for identifying violations must be consistent with the statement of the discharge limitation orders.

The third feature of monitoring that creates difficulties is that some pollutants are measured using "batch" or discrete sampling techniques.[36] This means that the choice of discharge limitation order and the source's optimal reaction to it should both be complicated by the choice of sampling regime (how often to sample and how many individual samples to draw at a time).[37] A fourth difficulty is that monitoring instruments are inevitably imprecise, that is, they measure with some error. This further complicates the task of defining and finding real violations.

Finally, all the above features of the monitoring problem take on a different cast when we drop the implicit assumption that sources will try to obey their discharge limitation orders. Cheating will be worthwhile if the possibility of detection and the penalty for a detected violation do not together provide a strong enough incentive. Where intermittent agency monitoring visits are involved, we further have to reckon with legal problems of access to sample, whether (and how much) advance notice is required, and how hard it is for the source to adjust discharges up and down—to avoid being caught cheating.

The above discussion was couched in terms of discharge limitation orders and violations as discharges that exceed these orders. A first point to note is that the monitoring problem is no easier or harder if what is involved is an emission charge where interest is clearly in an accurate periodic billing, and where a violation is, for example, a falsely stated emission total over some period. In fact, the same complications apply to the following three major forms of implementation system:

1. A nonmarketable permit stated in terms of allowed emissions (weight per unit time).
2. A marketable permit in the same terms.
3. An emission charge per unit of weight.

A second point is that we can imagine orders to sources in the outdoor air pollution case that will make a monitoring difference. For example, a requirement that a particular type of emission control equipment be installed is quite easy to monitor. That the equipment is correctly constructed and installed and can produce the desired discharge reduction is harder to determine. That it actually is achieving what it can achieve all or most of the time brings us back to the full-scale monitoring problem and is what we do, or at least should, care about.[38]

In the context of workplace air quality, monitoring involves similar statistical problems to those discussed above. That is, air quality will vary randomly; standards must take this into account; measurements are imperfect; and all these items complicate the identification of violations. One potentially vast difference in the workplace is the presence of the workers, who can in some cases carry personal dosimeters. These can either signal violations of an ambient air standard or could themselves be the basis of a standard. If the agency wanted an easier monitoring job and was willing to give up some amount of control of the actual air quality, it could opt for orders to firms specifying equipment and ventilation to be put in place. As observed above, it is much easier to monitor that such orders have been complied with than that air quality is what society has decided it wants. On the other hand, installation does not guarantee operation, much less proper operation.

Another type of implementation system, that of relying on liability for health damages, would involve its own unique monitoring difficulties. This is because of the need to know about individual exposure histories, at least to some extent. This, in principle, would involve following individuals after employment. Indeed, the system should have complete exposure histories available, which would mean complete employment histories for all employed individuals, and, more than that, complete histories of life-styles, habits, residences, and medical exposure to radiation and drugs. "Enforcement" of a liability system seems to amount to the same thing as applying the system at all. The difficulties are well known and very tough, involving burdens of proof, role of contributory negligence (or just plain living), and calculation of compensation.

When side-stream smoke is the problem, and the usual approaches of prohibition and zoning are in use, monitoring is fairly easy. Lighting up, or lighting up in the wrong area, is easy to spot, and a population of self-interested observer/monitors will exist in any public place. Enforcement, however, can be a problem. If the other occupants of the space have to do the enforcement by asking the violator to stop, the situation becomes fraught with psychological complexities involving fear of humiliation or violence, readiness of the observer crowd to become involved, and so forth. A potentially aggressive smoker in a setting of strangers may be able to get away with smoking openly. Officials who can intervene make a great difference.

Where a sales tax is used to discourage smoking generally, monitoring and enforcement have nothing to do with smoking. Rather, smuggling and a black market must be guarded against. This is too far afield for even this rather long paper. However, monitoring a fee per cigarette in particular settings would seem to be ridiculously difficult and expensive, so much so that such a scheme hardly deserves serious consideration.

Dynamic Characteristics: Incentives and Flexibility

This dimension should perhaps be split, for it has two aspects that display no neat relation to each other, being neither the same, nor opposites, parallel nor perpendicular. The first aspect is the incentive provided by the system for long-run adjustments by pollution sources and other actors. The second aspect is the flexibility of the system: the ease with which it adjusts to changes in tastes and technology, whether exogenous to society's pollution control efforts or induced by those efforts.

In the matter of incentive to technical change, the simple general rule may be summarized as follows: if compliance with an order is costly and if there is some choice of how to comply (what equipment or technique to use) then the source faced with the order will have an incentive to seek cheaper ways of complying in the long run. It is also true that for any particular source, an incentive system that puts a fee on the discharge remaining after control will create a greater incentive to change than will a regulation specifying that same level of discharge.[39] The possibilities for regulatory orders, however, include more than simple discharge limits. For example, orders that specify equipment or process to be used remove the incentive to change. More complex is the effect of the provisions of current pollution control legislation. In the Clean Water Act explicitly, and at least in the rhetoric surrounding the Clean Air Act, improvements in technology are supposed to trigger tightening of the standards.[40] This reduces the incentive to seek cost-reducing technical improvements in processes or treatment equipment, and under some circumstances may eliminate the incentive altogether.[41]

Other types of adjustment are possible as well, as have already been mentioned. Indeed, the existing more or less nationally uniform systems of air and water pollution control and occupational safety and health regulation are justified largely on the basis that they remove the possibility of adjustment in location. That is, firms cannot seek lower pollution control or safety costs through moving from one state to another. Of course, the ultimate adjustment is exit—shutting up shop to avoid the cost—and this cannot be prevented in a free economy, though rules about consultation and compensation can make exit less attractive by raising its price.[42]

There is another layer of adjustments as well: that done by consumers who face changing relative prices due to differences in pollution control costs. It is difficult to make quantitative predictions about the ultimate effects of these adjustments because they involve the nature of the demand for specific goods as functions of their own prices, of consumer incomes, and of the prices of other goods, all of which will be changing at the same time as pollution control or safety regulations are implemented. But this consideration does bring us back to smoking. Any regulation or incentive that applies to all smoking and raises the cost of smoking relative to abstaining will tend to reduce the overall incidence of smoking and the total amount of tobacco smoked. Because of the habitual nature of the act, the personal analogs of the cost-of-discharge-reduction functions may exhibit peculiar properties such as threshold effects and other discontinuities, making prediction of long-run effects of policy very difficult. And because standards (limits on smoking generally or on cigarettes per person per hour in particular places) seem ruled out by monitoring difficulties, we do not have here a neat contrast between regulations and incentives. Increasing the price of a package of cigarettes may or may not have a greater effect on long-run behavior than segregating smokers or prohibiting smoking in public places.

This last set of observations does, however, raise an issue related to the next section on ethical concerns. It is not only the regulations or charges that influence long-run behavior. The messages sent about how society judges particular forms of behavior are also important. These may be explicit, as in the application of criminal sanctions in the pollution control acts or in advertising campaigns designed to discourage smoking.[43] One might view our current approach as including a set of regulations and adjustment rules that tend to discourage pollution control and safety innovation, along with a propaganda campaign designed to encourage that innovation, all backed up by publicly funded research. This internally inconsistent combination of policies strikes many as a high price to pay for the inclusion of stigmatizing language in the statutes.

Technology and tastes also change for reasons having nothing to do with pollution control, occupational health and safety regulation, or campaigns to control side-stream smoke. New products appear, new resources become available, and people both react to the changing menu and somehow generate their own fads, fashions, and trends. As these changes are reflected in the size, distribution, and character of industrial and government activity, the nature of the air pollution control problem changes in each of the settings. The second important dynamic aspect of an implementation system is its flexibility in the face of these inevitable changes.[44]

It is first necessary to be clear about what counts as flexibility. I shall use that word to mean the ease with which the system maintains the desired ambient standards as the economy changes. The most important measures of ease are in turn: first, the amount of information the agency has to have and the amount of calculation it has to do to produce the appropriate set of orders or incentives for a new situation; and, second, the extent to which adjustments involve a return to a politically sensitive decision-making process.

An initial observation is that this dynamic aspect of implementation is likely to be a concern principally in the outside air pollution setting (so long as my assumption of fixed ambient standards is allowed to stand). In the workplace setting, the presence of the single decision maker (or source) means that translating changed conditions into a fixed ambient standard is unambiguously that source's problem. A factory may be expanded, new production processes installed, new inputs used. The owner has to decide how to remain in compliance.

In the side-stream smoke problem, discussion of the efficiency and monitoring dimensions led us to conclude, at least tentatively, that maintaining a uniform ambient standard in public places almost certainly could not be done through regulations or incentives attached to individuals and their smoking decisions. Either a ventilation rule would be necessary, in which case the context would become rather like the workplace, or segregation or prohibition would have to be instituted. The latter implementation systems can be thought of as constituting the policy goals in themselves. That is, an ambient standard no longer drives the system in any way. Whatever ambient smoke levels result from segregation (or the pristine result of prohibition) are accepted. Changing circumstances make no difference. If a ventilation standard is the implementation system chosen then flexibility is an issue, for changing average smoking habits might well imply a change in the required level of air exchange to maintain a particular standard. The calculation of that change in requirements would have to be done at the center, that is, in the regulatory agency, implying a significant information and calculation load. In addition, revising and reissuing regulations covering thousands of public places with private owners would inevitably stimulate at least attempts at political maneuvering to change the effective standard.

In the outside air pollution setting the problem of flexibility is more complex, if only because the range of implementation systems that must be taken seriously is very much larger. Because of the number of actors affecting the ambient concentrations, only one implementation system is, even in principle, completely flexible (requiring no agency intervention and hence no information, calculation, or political activity). This system is one based on marketable ambient air quality permits. Because those permits (if appropriately enforced, of course) automatically preserve the desired ambient quality, no agency action is required as the economy changes; private permit trades take care of desirable adjustments. (Enforcement, as already noted, does require information in the form of dispersion models for translating ambient permits into discharge limits.)

No other system of outdoor air pollution control can maintain an ambient standard as conditions change, at least not without adding additional constraints. The common regulatory approaches, involving nonmarketable permits stating allowed discharges, protect ambient quality only in the absence of new sources. The U.S. system with discharge permits for existing sources and different, more stringent rules to define the permit terms for new sources, would have protected ambient quality by imposing what would have amounted to a flat prohibition of new sources where ambient standards were already violated. The offset system was EPA's response to this politically explosive possibility. Offsets allow new sources to buy pollution rights from existing sources and thus make the system more like a marketable permit approach. The multisource bubble (the assumption that what matters is total discharge from two or more stacks contained under an imaginary bubble), another EPA innovation, is essentially a trade among existing sources. Together, and with the accompanying restrictions on who can trade with whom and how offset values change as the traders are further apart or closer together, these policies are roughly equivalent to a marketable discharge permit scheme.[45] Such a scheme can adjust to change, but in the process can produce violations of ambient quality. If ambient quality is absolutely protected, then either the ability to adjust must be restricted and calculations by the regulator of allowable trades required, or the initial permit total must be very low and the whole scheme relatively expensive and restrictive.

Emission charges do not protect ambient quality unless they are adjusted as change occurs. Such adjustment requires new calculations at the agency if the charges are to be efficient. (And then, because the charges are individually tailored, each change is a fresh chance for political action.) If the charges are uniform and set by trial and error, adjustment will involve the expense of error, and static efficiency will not be achieved.

Thus, overall, the two dynamic considerations push toward marketable permit systems for outside air pollution control and do not have much to say to us in the other two settings.

Political and Ethical Appeal

Discussing political and ethical appeal is a tricky business. One obvious problem is the tendency to add to the question, is this system politically appealing?, the words, "to me." A second is that, as pointed out in the introduction, arguments about political appeal have a self-fulfilling character. Voters and even their representatives are not well equipped to judge how a particular implementation system has affected them, and even less well equipped to judge that system against another that has not been adopted. Arguments about political appeal, however weakly based, can thus substitute for reality. Ethical appeal does have an analytical life of its own, but in its practical manifestation shows up as a part of political appeal.[46]

My plan for attempting to tread this mine field is to begin with the assumption that the political appeal of an implementation system for pollution control has two major determinants: ethical appeal and distributional character. In assessing ethical appeal, I shall in turn concentrate on four issues:

1. How the system implicitly or explicitly judges polluting behavior—as wrong or as neutral.
2. To what extent the system allows the polluter to choose actions and, specifically, to choose actions inconsistent with the achievement of the social goal (ambient standard).
3. How great a risk of failure (to achieve the social goal) the system carries and, just as important, whence arises that risk.
4. What motives the system plays on to manipulate polluter actions.

As a way into the problem it will be convenient to examine a situation in which we know something about political appeal (or lack thereof) and about the ethical and distributional qualities that lie behind it. Thus, it is widely agreed that emission charges are not a politically attractive way to achieve ambient quality goals. This is attested by the repeated failures of the Congress and state legislatures to enact bills incorporating them, and by the failure of the few places (such as Vermont) that have created such systems in law to put them into practice.[47] While no one can be entirely sure why charges have failed to appeal, it is easy to identify a number of distributional problems and ethical objections that have been adduced as reasons. From this base of observation it will be possible to branch out in two ways. First, a critique of the ethical objections to charges will indicate to what extent the alternative systems are really different on this dimension. Second, discussion of the workplace and side-stream smoke settings will reveal a few interesting contrasts.

Emission charges lack political appeal. That much is clear. Some of the reasons for this failing may be inferred from the objective characteristics of that implementation system. Other, principally ethical, objections have been stated explicitly by opponents and policy analysts.

A potentially serious distributional problem does arise from the operation of the charge. Because it applies a price to remaining discharges after the application of control technology, it costs each discharger more than would an order requiring the installation of that technology.[48] This payment is the continuing incentive to seek better technologies and therefore is responsible for the desirable dynamic or long-run characteristics of the charge. But to the polluter, it is an out-of-pocket cost that might equal or exceed the cost of treatment.[49] Thus, any efficiency gains (the size of which depends on how the charge is structured, as was demonstrated above) are purchased at great distributional cost from the polluters' point of view. And while it is possible that the gains might be great enough to outweigh the charge payments, which are in any case "mere transfers," it is unlikely that a system could or would be adopted to leave the polluters no worse off than under regulation while not disturbing the incentive structure.[50]

A second distributional objection sometimes raised to charges is that the "rich" can pay and pollute while the "poor" must comply. But this point applies only to charges levied on individuals in relation to activities that directly give pleasure—the excise tax on cigarettes, for example. Firms are not rich or poor in the same sense as individuals. Nor is it usually assumed that the production activities of the firm are pleasurable per se for the management or owners. The purpose of the firm is to make a profit, and a perverse reaction to a charge will reduce profit relative to the outcome from a rational response.

It is true that the charge payment added to the resource cost of treatment could be too much for some firms tottering on the edge of insolvency even before the imposition of the pollution control policy. In this sense a charge system could be said to be harder on firms in poor financial shape than a regulatory system aimed at the same result. But this is just another version of the first objection; not an instance of the rich being able to pay to pollute.

To the extent this ability-to-pay argument does have force, it would seem to be in cases of monopoly or regulation where cost pass-throughs are more or less automatic. Then the firm's profit is protected no matter how rational or irrational its adjustment to the charge.

On the explicitly ethical side the idea of an emission charge has been widely criticized.[51] By making pollution into an activity that one can pay to engage in, the charge approach implicitly judges it to be ethically neutral—like buying a shirt. Those who judge polluting activity to be wrong in itself, as a crime against other persons or against nature, do not approve of that result. They want the law to stress absolute prohibitions and wrongness.[52] Second, and a related objection, the charge allows the polluter a choice of how to behave. Not only is choice undesirable because of the wrongness of the act, but that choice leaves room for behavior that can prevent attainment of the social goal. Thus, the charge approach risks failure in an obvious way. Finally, to the extent that polluters do choose the courses of action consistent with attainment of the goal, they do so on the basis of a bad motive: the profit motive. It would be best if they chose the right way for the right reason: a sense of social duty or whatever one wants to call it. It even appears proponents of this line of criticism would prefer to see compliance compelled by fear than to see it induced by the lure of profit.

If emission charges fail each of these tests, what can be said of the alternatives; and what of the tests themselves? The two distributional problems have already been dealt with. The first—the transfer from polluters—is not a problem with regulatory orders. The second—the rich can pay to play—is again a valid objection only in contexts involving individuals and activities that directly give pleasure.[53] The first ethical objection, to the ethically neutral message of the charge, must also be placed against the substance of any reasonable system of regulatory orders. That is, since zero pollution levels are not in fact attainable, reasonable regulations such as the ones we live with today will specify acceptable nonzero levels of discharge. Such specifications seem to me to carry the same message as would a charge. What carries the ethical burden in current legislation is the rhetoric in the statutes and their legislative histories. Most obviously, the zero-discharge goal of the Clean Water Act is a statement implying that no discharge is acceptable in the long run. Such rhetorical flourishes are easily transferable and could just as well—and to just as much practical effect—be tacked onto a charge-based implementation system.

On the related matters of choice, risk, and motive, the beauty of regulations is in the eye of the beholder. It is true that under a system of discharge limitation orders, a discharger will be in violation if it chooses to discharge any amount over that specified in the applicable permit. Under a charge system, no violation is implied by any level of discharge. In this way, the regulatory system reduces the choice to a binary one: comply or do not comply. This reduction in the range of permitted behavior—the fact that discharge over the agency's target for a particular source is treated as a violation of law—may reasonably be seen as reducing the chance that undesirably high discharges will in fact be chosen. But it is very hard to judge how much of a reduction in this chance we can expect. It can certainly not be said that regulations reduce the range of choice to zero, for the choice to violate permit terms is always open. The section on monitoring and enforcement stressed both how hard it is to find out whether sources really are complying and how similar the monitoring problem is for charge and limitation order. Further, enforcement as a practical matter involves monetary penalties for noncompliance. To be credible these must be related in some way to the cost savings from noncompliance. (In the current Clean Air Act, the relation is meant to be explicit and exact.) Therefore, only to the extent that criminal sanctions are included, and that public opprobrium can be attached to permit violations, would the regulatory system seem a priori to have a significant edge on this risk dimension. Finally, as just pointed out, the motive for obeying a discharge limitation order must be at least partly pecuniary, for the penalties are at least partly in this form. The actual mix of civil and criminal enforcement will determine how compliance is viewed by dischargers. The mere existence in the statute of criminal sanctions does not settle the question of motive.

The problem of managing workplace air quality seems quite similar to the outdoor pollution setting when we are considering political and ethical questions. The transfers created by charges on polluting behavior could be substantial and disruptive. The argument about the rich paying to pollute would be similarly inapplicable. And the above comments about tone, choice, risk, and motive all seem applicable. A different set of questions would arise if we intended to depend on a liability system or some variation of workmen's compensation to produce compliance with the social goal. Then, the difficulty of proving the cause of chronic diseases would seem to introduce a very much larger risk of failure. Further, we might very well judge there to be something ethically undesirable about a system that involved an incentive for the firm to marshal its legal and scientific might against sick workers asserting a claim that workplace air pollution was a cause of their sickness.

The side-stream smoke setting exhibits two important differences from the other two. First, an activity that gives pleasure directly to individuals is involved. Thus, any implementation system based on a tax (per pack or per cigarette lit in a particular place) runs squarely aground on the distributional objection that the rich can pay to pollute. In other words, the burden of achieving the social goal would be shifted to the less well off rather than shared on some appealing basis such as the initial contribution to the problem. Second, again because individuals and not firms are directly the objects of policy, attempts to stigmatize behavior, through statutory rhetoric, criminal sanctions, and even advertising campaigns, might be expected to yield substantial results. The owners or managers of a firm, told that pollution is wrong, can argue for their own and society's benefit that production is right and that some waste is inevitable: that the policy argument should be about amount and form. The individual, told that smoking in public places is wrong, has no such counterarguments. No social good is served by the activity, and it is certainly not an inevitable result of any activity that does produce a social good. Thus, the claim that a regulatory, even an exhortatory, public policy is to be preferred on ethical grounds may at least not be dismissed in the side-stream smoke setting.

Some Concluding Comments

The conclusion of this chapter cannot be a neat and decisive one. For one thing, we have found that none of the contending implementation systems obviously dominates in any of the settings on all the dimensions of judgment. Thus, a choice between methods in any setting requires a weighing of the relative importance of each dimension; in effect, a further social decision problem. We have even found that along single dimensions there is often little to choose between the contenders. The standard arguments in the literature of environmental policy analysis, arguments that stress vast differences on efficiency, monitoring, or ethical grounds, were found to be sometimes wrong, sometimes misleading, and sometimes incomplete. For example, the static efficiency attributes of charges have been oversold because proponents have neglected the necessary complexity, inflexibility, and information intensity of efficient charge systems. The monitoring disadvantages of charges have, on the other hand, been overstated by those who did not recognize the substantial difficulty of monitoring for compliance with time-averaged discharge permits. Ethical objections to charges in turn depend for their validity on a view of pollution as wrong, a view which would seem to rule out any reasonable regulatory system as well.

In my own judgment what we are left with may be summarized briefly as follows:

For outdoor air pollution, static efficiency is an expensive goal when we take information load and administrative complexity into account. But the desirability of clear, unambiguous dynamic incentives and of flexibility in the face of exogenous economic change point to marketable discharge permits as the most promising implementation system. One or

258 To Breathe Freely

another twist may be added to the simplest system to reduce the chance of spatial "hot spots."

For workplace air quality, the setting of ambient standards is a direct method of pursuing the social goal. It should call forth statically efficient behavior, be as easy as any competitor to monitor and enforce, and display no obvious political or ethical defects. The incentive it creates for long-term change in production technology is perhaps not as great as a charge would introduce. But it is not without such an incentive and overall seems a winning choice. Liability and insurance systems seem to pose too great a risk of failure to be acceptable as the primary implementation system.

For side-stream smoke in public places, no approach attempting to regulate "discharges" at nonzero levels, or based on a charge per pack or per cigarette smoked in particular settings, seems to have any practical value. This is because of monitoring and enforcement problems not offset by clear efficiency gains; and, for any charge, because the rich could pay and shift the compliance burden to the poor. In this setting spatial segregation and outright prohibition seem both more practical and ethically preferable. The ventilation requirement might be feasible but is inflexible in the face of change and sends no message to smokers.

Notes

1. The problem is further complicated when we consider the stochastic nature of wind and weather. Predictions, in fact, must involve the probability that the ambient pollution concentration at some point will be less than the goal, or the expected amount of time the goal will be met.

2. For now let us speak of "orders"—postponing the central discussion of the chapter—whether in particular situations orders are in fact to be preferred to inducements or incentives that leave the choice of discharge level to the discharger.

3. Another possibility for identifying at least one member of the technically feasible set is trial and error, that is, for example, setting a specific discharge limit for each pollution source in the region, observing the resulting ambient quality, and adjusting the limits down if the goal is not met or relaxing them if the goal is overmet. Such an approach has been advocated and the problems with it will be discussed below.

4. The larger problem, choosing a preferred strategy with no accepted single measuring rod, is the subject of the chapter and is definitely not a technical problem. Rather, it is a political one.

5. It is worth pointing out that here and in what follows feasible combinations of discharge orders are those implying that the desired ambient standard is not violated at any monitoring point. The air quality at any or all points may be better than the goal. But if the quality achieved is better than the goal at every point, there exists any number of just feasible points that cost less and result from reducing quality at one or more points just down to the announced goal. If one both takes the social choice of the desired ambient goal seriously and believes that the costs of discharge control are good only insofar as they produce higher quality and not good in themselves, then the order combinations that produce super-good quality are of no interest.

6. For political purposes, initial incidence is usually very important, though more or less serious efforts may be made to go beyond the first round and look at how and to whom costs are passed on.

7. In each of these cases the measure of "profit," the range of possible adjustments, and the nature of the decision process will be different. Thus, for a given implementation system, the adjustments of the three broad classes of actor may be quite different.

8. Even the regional pollution dispersion mechanism might change with changing types, levels, and locations of human activity, as with shifting urban "heat islands," e.g., William R. Frisken, *The Atmospheric Environment* (Washington, D.C.: Resources for the Future, 1973).

9. As already discussed, by changing residence, job, asset portfolio, or habits, individuals can change their own net benefits from a particular policy (goal plus method of accomplishment).

10. Because neither costs nor benefits are easy for individuals to determine in the air pollution area, most individuals may find it hard to judge where their self-interest stands once any option is operating. Thus, predictions may create opposition that would not otherwise exist. Indeed, the idea that opposition ought to exist may itself be enough to do in a plan.

11. This view is reported and justified in Steven Kelman, *What Price Incentives*? (Boston: Auburn House Publishing Co., 1981).

12. This view is explicit in much of the work of my colleague Allen V. Kneese; for example: Allen V. Kneese, Robert U. Ayres, and Ralph C. d'Arge, *Economics and the Environment* (Washington, D.C.: Resources for the Future, 1970).

13. A recent National Academy report on risk, for example, says, "Coping methods [implementation systems] that preserve or increase free choice by individuals should be considered." Committee on Risk and Decision Making, *Risk and Decision Making: Perspectives and Research* (Washington, D.C.: National Academy Press, 1982), p. 23. However, it seems likely that those who see pollution as wrong in itself would quarrel with the very notion of collectively choosing a nonzero level of pollution and calling it acceptable. The only acceptable goal would be for them the elimination of pollution. This is only one of the more obvious points at which the analysis of this chapter rests on implicitly assumed ethical positions.

14. Of course every system, even in a centrally planned economy where there are no autonomous firms, must offer some choice; for even the most explicit rules backed by the most severe penalties may be disobeyed. The question is one of the relative cost of perverse choice.

15. What qualifies as opportunity and ability depends on the definition of the air quality standard and the ease of identifying violations. Thus, to monitor a 24-hour outside pollution standard for, say, SO_2, requires a monitor in place for over 24 hours and sophisticated measuring devices. Monitoring a side-stream smoke standard that was qualitatively defined for every instant in time would require very short "residence time" and only a sense of sight or smell. Between these two extremes fall, for example, eight-hour-averaged workplace air standards for which portable personal dosimeters are available.

16. A simple illustration of the need for individually tailored orders or charges is provided in the appendix at the end of this chapter.

17. If either there is only one source affecting the region, or if all discharges are perfectly mixed together in the atmosphere and produce a uniform ambient concentration, then a uniform charge can be used and trial and error might work. Nearly perfect mixing may characterize conditions during atmospheric inversions in which regional ventilation is poor. This is often referred to as the "mixing-bowl" situation.

18. Trial and error is impractical because of the large number of "knobs" available to twist in a multisource region. If each of only ten sources could control to each of only three levels of discharge, there would be over 59,000 possibilities for an initial trial. That first trial might eliminate some fraction of the options as either infeasible or unnecessarily strict, but homing in on the one best option might easily involve 10 or 20 very expensive trials.

19. Because in general we cannot sort out the influence of individual dischargers on measurements at a point in the environment, an ambient pollution permit system must be tied to discharges by the same kind of regional atmospheric model found necessary for determining efficient orders and charges. This tie is necessary for two purposes: to allow dischargers to determine what permits they must hold for any discharge level and to allow the regulatory agency to monitor source emissions to see if they are consistent with the portfolio of ambient permits held.

20. This is illustrated in the appendix at the end of this chapter.

21. The situations are the same because the uniform charge and the uniform market price play exactly the same role in the calculations of (rational) dischargers.

22. An additional possibility, coinciding roughly with the ad hoc system now being cobbled together at EPA, begins with existing permit levels and does not actually require either cost or atmospheric information. But then it does not protect ambient quality against the results of trading either.

23. For a proof see David Montgomery, "Markets in Licenses and Efficient Pollution Control Programs," *Journal of Economic Theory* (December 1972): 395–418.

24. When adjustments are taken into account, at least a limited case can be made for a charge in the workplace setting, since such an approach would supply a continuing incentive for the firm to improve workplace air quality.

25. A combined system can also be imagined, involving no-fault compensation unless there was a showing of willful violation or gross negligence, in which case tort or even criminal liability would apply.

26. The cast of receptors will also be constantly changing, which means that a single ambient standard is itself likely to be unsatisfactory.

27. There is also a distributional, hence political, problem with charging a fee to smoke. See below.

28. Such a requirement, since it would be placed on the owner/operator of the public space, would make the side-stream smoke context look rather like the workplace air quality problem. A single "source" (in this case not a source of pollution at all but the source of ventilation) would replace the multiple sources. An ambient standard would apply almost everywhere—everywhere, that is, but within a few inches of each smoker.

29. Harrington found from New Mexico data that stationary source air pollution sources were in violation of their emission standards about 40 percent of the time on an average over a number of sources. Winston Harrington, *The Regulatory Approach to Air Quality Management: A Case Study of New Mexico* (Washington, D.C.: Resources for the Future, 1981), pp. 104–8.

A rather large study of compliance by stationary sources of air pollution done for EPA by a group of consulting firms presents its finding in a different form and is a bit hard to interpret. It appears, however, that based on quantitative data and qualitative (inspection) information, the sample of sources was out of compliance about 10 percent of the time. Measuring compliance by emissions, the same group found total excessive emissions about 25 percent above the allowed total. But because the average source was "overcontrolling" for the time it was in compliance, the actual annual emission totals were estimated to be less than the implied annual allowance. The report does not make clear the terms of the permits applicable to the several plants, but one would expect them to involve allowed rates (e.g., lbs. of particulates per 10^6 Btu or per ton output), or perhaps quantities emitted per day. It is unlikely, therefore, that the annual average results are a useful indicator of compliance. Robert C. McInnes and Peter H. Anderson, *Characterization of Air Pollution Control Equipment Operation and Maintenance Problems* (Washington, D.C.: USEPA, 1981), esp. p. iv.

30. Section 120 of the Clean Air Act is devoted to a noncompliance penalty system. See also William Drayton, "Economic Law Enforcement," *Harvard Environmental Law Review* 4, no. 1 (1980): 1–40.

31. Self-monitoring is a standard requirement of air pollution permits written under the New Source Performance Standards (NSPS) for power plants. Returns from a survey of state agency monitoring practice, which we conducted at RFF, suggested self-monitoring is also a widespread requirement for existing sources, especially large sources, of both air and water pollution discharges.

32. EPA's latest regulatory reform idea is to make the discharger himself the nominee. This further complicates, but hardly can be said to solve, the problems discussed in this paragraph and the one that follows. See, for example: *Air/Water Pollution Report*, "Corporate Environmental Auditing Concept Is Still Being Considered by EPA," July 12, 1982, p. 271 and Joseph F. Guida, "A Practical Look at Environmental Audits," *Journal of the Air Pollution Control Association* 32 (May 1982): 568–73.

33. This statement must be qualified in two ways. Remote monitoring equipment makes it possible to measure concentrations of certain residuals in a stack plume, though these

methods are neither simple nor precise. (See M. R. Williamson, "SO_2 and NO_2 Mass Emissions Surveys: An Application of Remote Sensing," in *Continuous Emission Monitoring Design, Operation and Experience* (Pittsburgh: Air Pollution Control Association, 1981), pp. 303-7. Somewhat more tenuous is the technique of using ambient quality levels and discharge composition to infer discharges, though it might in some cases provide a defensible check on self-monitoring data. (See F. E. Courtney, C. W. Frank, and J. M. Powell, "Integration of Modeling, Monitoring, and Laboratory Observation to Determine Reasons for Air Quality Violations," *Environmental Monitoring and Assessment* 1, no. 2 (1981): 107-18; and Glen E. Gordon, "Receptor Models," *Environmental Science and Technology* 14 (July 1980): 792-800.

34. Some individuals may be motivated by bounties or simple outrage to report on obvious offenses, but in the current state of law and technology they cannot be counted on to do more than spot the tip of the iceberg.

35. At its simplest this means that if the agency orders a source to hold its discharges below D at all times, the source must actually aim at a target or mean discharge value far enough below D that random occurrences of excess emissions will be so infrequent as to be ignored. How far below D the target emission must be depends on the width of possible swings in discharge, the costs of control, and the penalties for detected violations. If the regulatory agency wants to see the source emit D on average, it must redefine a violation. For example, if it knew the distribution of actual discharges around the source's target, it could define a violation as, for example, any discharge greater than $D + K$. K would reflect how closely the source could control its emissions and would be matched to an appropriate penalty reflecting the costs of this control and the acceptable probability of really high emissions (greater than $D + K$).

36. It appears that continuous sampling methods with automatic recording are being developed for more and more pollution types, so this difficulty may tend to disappear as time goes on.

37. Sampling size and frequency, given the source's distribution of discharges and the characteristics of the tests performed, define the probabilities of missing the violations and of finding false violations. William J. Vaughan and Clifford S. Russell, "Monitoring Point Sources of Pollution: Answers and More Questions from Statistical Quality Control," *The American Statistician* 37, no. 4 (November 1983).

38. It is worth stressing here that without some orders or incentives applying to specific sources, monitoring could not lead to practical, legally sustainable enforcement. Thus, if our air pollution control system involved only an ambient standard, monitoring evidence of its violation would only mean that the sources in aggregate (and as located) were emitting too much. There would be no basis for taking any particular source to task.

39. This is easy to show. In the figure below, the firm's initial marginal cost-of-discharge-reduction curve is MC_0. Assume it is complying with an order to discharge no more than D_0. This could also be achieved by the agency charging a fee of e_0 per unit of discharge. The order costs the firm area A, the cost of control to D_0. The charge would cost it area $A + B$, the control cost plus the total fee paid on remaining discharges. If, as shown in the second panel, the firm can find a way to reduce its costs to MC_1, it saves C under the order system and $C + G$ under the charge. The new discharge, D_1, under the charge system is lower as well. This result also applies to marketable permits, for the permit price corresponds to the charge even though it may not be paid out of pocket by the originally permitted sources.

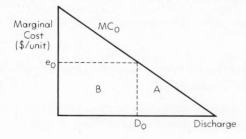
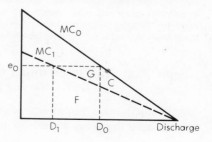

40. Clean Water Act, Section 302 d in *Environmental Statutes* 1980 (Washington, D.C.: Government Institute, 1980).

41. A very simple way of looking at this process uses the figure in Note 39. When technology is improved, and marginal cost falls to MC_1, the ratchetting-down requirement implies a new lower discharge standard. Let us say that the rule for choosing this level is to maintain equal marginal costs (e_0) before and after. Then after technical change, the standard would be D_1, and the net savings to the firm would be $C - F$. In this figure, area F will always be greater than area C, so there is a disincentive to innovate. More generally, the existence of the additional cost, F, will at least reduce the positive incentive to innovate.

42. For an interesting account of just how expensive exit has gotten for U.S. business, see Thomas F. O'Boyle, "High Cost of Liquidation Keeping Some Money-losing Plants Open," *Wall Street Journal*, November 29, 1982, p. 29.

43. For a related experience, see E. Lewit, D. Coate, and M. Grossman, "The Effects of Government Regulation on Teenage Smoking," *Journal of Law and Economics* 24 (December 1981): 545–76.

44. I am not speaking here of changes in the ambient standards due to advances in scientific knowledge or changing (possibly implicit) social valuation of human life. Rather I am assuming fixed ambient standards embedded in a moving economy.

45. "Roughly" especially because new sources still face technologically based maximum discharge requirements.

46. For analysis of relevant ethical systems and judgments by persons with expertise in this area, see Chapters 4, 5, 6, and 7, this volume.

47. Charges on wastewater discharges are fairly common in Europe. (See, for example, Ralph Johnson and Gardner Brown, *Cleaning Up Europe's Waters* [New York: Praeger, 1976].) But with the exception of the new German national charge scheme these systems are not really incentive systems. Rather they are designed to raise revenues for treatment plant subsidies and for other investments intended to improve water quality. See Blair T. Bower, Rémi Barré, Jochen Kühner, and Clifford S. Russell, *Incentives in Water Quality Management: France and the Rhur Area* (Washington, D.C.: Resources for the Future, 1981).

48. For example, on potential steel mill costs and charge payments, see Clifford S. Russell and William J. Vaughan, *Steel Production: Products, Processes, Residuals* (Baltimore: Johns Hopkins University Press for RFF, 1976).

50. A system attempting to get around this problem without destroying the incentive effects of the charge has been invented by James D. Smith. The complexity of Smith's solution seems sufficient to prevent its serious consideration, though it may signal a useful line of inquiry. See J. D. Smith, "Feasibility Study: A Fee/Subsidy System for Controlling Sulfur Dioxide Emissions in Philadelphia," City of Philadelphia Air Management Services, September 1981.

51. I am indebted to Steven Kelman for his discussion of the ethical objections to emission charge systems. Kelman, *What Price Incentives?*

52. Some of these objectors use phrases indicating that using a charge somehow "leaves polluting behavior up to a market." In fact, of course, charges are administratively set by government to achieve a goal; they are not determined by the decentralized interplay of demanders and suppliers. The society retains control in a charge scheme, unless one takes seriously the idea of widespread perverse behavior by polluters.

53. Note that not even all consumption activities involve direct purchase of pleasure. We buy fuel oil to produce heat and hot water in our homes, but we derive no direct pleasure from burning the fuel oil. Therefore a differential tax tied to fuel oil sulfur content could not validly be objected to on the-rich-can-pay grounds. Neither rich nor poor would gain anything from paying more to burn higher sulfur oil.

Appendix: Simple Demonstrations of Propositions Concerning Charges and Standards

Let us discuss the management of outdoor air quality in a very simple region where there are only two sources of pollution (say SO_2), A and B, and a single monitoring station at which the desired ambient standard is defined. A is located upwind of the monitoring station, and one ton per day of discharge from A translates into a contribution of one microgram per cubic meter ($\mu g/m^3$) average 24-hour concentration at that point. B's location is crosswind, and its contribution to monitoring point concentrations is only 0.25 $\mu g/m^3$/daily ton of discharge.

In the precontrol situation, A and B are both emitting 100 tons/day (T/d), and together produce a concentration at the monitoring point of 125 $\mu g/m^3$. Society has, however, decided to define an ambient standard of 75 $\mu g/m^3$ at the monitoring point.

Finally, the agency knows the marginal cost of reduction functions for the two dischargers. Let MC_A be the marginal cost at A and R_A be the reduction achieved at A, and similarly for B. Assume we know that:

$$MC_A = 1.5\ R_A$$
$$MC_B = 0.75\ R_B$$

Notice also that $100 - R_A$ is the remaining discharge at A (D_A) and similarly for B.

First, consider what total regional discharge could be allowed if we did not know in advance which source(s) would discharge it. This quantity would be 75 T/d, which, if discharged by A, would just meet the standard. There would be any number of ways to go from the precontrol total of 200 T/d to 75 T/d, but in order to allow comparisons with later results let us seek the cheapest combination of initial discharge limitation orders. (Recall, however, that the purpose of using the 75 limit would be to avoid exactly this kind of calculation. The agency might just issue marketable permits totaling 75 T/d in the same ratio as original discharges: 37.5 to A and the same to B, counting on the "market" to produce the final allocation).

To solve this problem in this simple context it is sufficient to use a technique from calculus involving the "Lagrangian multiplier." This is a trick allowing us to write the constraint, in this case the limit on total discharges, in one expression with the costs to be minimized. In a more complex case, with many sources and monitoring points, we would have to use a so-called "programming" method that allowed for the possibility that the cheapest solution might involve better quality than 75 $\mu g/m^3$ at some monitoring points.

For this example, the Lagrangian expression is:

$$L = \frac{1.5 R_A^2}{2} + \frac{0.75 R_B^2}{2} - \lambda\ [(100 - R_A) + (100 - R_B) - 75]$$

where the first two terms are the total reduction costs at the two sources, λ is the Lagrange multiplier, and the expression in brackets is the requirement that the postpolicy discharges equal 75 T/d in total. Solving for the cheapest policy involves partial differentiation of L with respect to the "policy variables," R_A and R_B, and λ. The results of this exercise, the so-called "first-order conditions," are:

$$1.5\ R_A + \lambda = 0$$
$$0.75\ R_B + \lambda = 0$$
$$100 - R_A + 100 - R_B = 75.$$

Solving this set of simultaneous equations yields the cheapest combination of R_A and R_B:

$$R_A = 41.67$$
$$R_B = 83.33$$

implying initial discharges:

$$D_A = 58.33$$
$$D_B = 16.67$$
$$\text{Total} = 75.00.$$

The cost of this policy would be

$$0.75\ (R_A)^2 + 0.375\ (R_B)^2 = 1302 + 2604 = 3906.$$

Notice, by the way, that a distribution requiring equal reductions at both sources would be even more expensive. Let $R_A = R_B = 62.5$, so that $D_A = D_B = 37.5$.

Then total cost $= 0.75\ (62.5)^2 + 0.375\ (62.5)^2 = 4395.$

Consider now another possible solution to the implementation problem. This is the uniform emission charge (or, alternatively, the tradable discharge permit when we do have an idea what the final trading result will be). Here we can take advantage of the uniformity of the emission charge (call it e) to write immediately a set of requirements that the solution must satisfy:

$$1.5\ R_A = e = 0.75 R_B$$
$$100 - R_A + 0.25\ (100 - R_B) = 75.$$

To which the solution is:

$$R_A = 33.33 \text{ so } D_A = 66.67$$
$$R_B = 66.67 \quad D_B = 33.33$$
$$\text{Total} = 100.00.$$

The total cost, without charge payments, is:

$$0.75 (33.33)^2 + 0.375 (66.67)^2 = 833 + 1667 = 2500.$$

The ambient quality standard is met:

$$D_A + 0.25 (D_B) = 66.67 + 0.25 (33.33) = 75$$

Note that the unit charge e would be:

$$1.5 (R_A) = 0.75 (R_B) = 50$$

and the total charge payments would be

for A: 66.67 × 50 = 3333.5
for B: 33.33 × 50 = 1666.5
                   5000.0

or twice the resource cost of discharge reduction.

Finally, let us find the efficient solution: the lowest cost combination of discharge orders, or the corresponding emission charge(s). The Lagrangian expression is:

$$L = \frac{1.5 R_A^2}{2} + \frac{0.75 R_B^2}{2} - \lambda [(100 - R_A) + 0.25 (100 - R_B) - 75]$$

The first order conditions are:

$$1.5\ R_A + \lambda = 0$$
$$0.75\ R_B + 0.25 \lambda = 0$$
$$100 - R_A + 0.25 (100 - R_B) = 75.$$

To which the solution is:

$$R_A = 44.44 \quad \text{so}\ D_A = 55.56$$
$$R_B = 22.24 \quad \quad D_B = 77.76$$
$$\quad\quad\quad\quad\quad\quad\text{Total} = 133.32$$

Check for the ambient quality:

$$D_A + 0.25\ D_B = 55.56 + 0.25 (77.76) = 75$$

Total cost:

$$0.75 (44.44)^2 + 0.375 (22.24)^2 = 1481 + 185 = 1666$$

or 67 percent of the uniform charge total cost, and only 37.9 percent of the least information-intensive solution. This solution could be induced by imposing the following nonuniform emission charges:

Unit charge at A = 1.5 (44.44) = 66.66
Unit charge at B = 0.75 (22.24) = 16.68.

The charges would produce the following total charge payments:

By A: 66.66 × 55.56 = 3704
By B: 16.68 × 77.76 = 1297
    Total payments = 5001.

Epilogue: Hazards of Entering the Risk Debate

13

On Not Hitting the Tar-Baby: Risk Assessment and Conservatism

Langdon Winner

Contemporary discussions of risk contain a fruitful possibility. Addressing ways in which the broader effects of industrial production can damage environmental quality and endanger people's health and safety, such inquiries approach one of the most important questions that face modern society: are there justifiable limits to the application of today's scientific technology? Risk assessment goes beyond vague musings about human values in a technological age to talk about issues that are at once concrete and normative. It seems to offer a way to act upon one's sense of moral responsibility at a time in which progress in material culture no longer has an obvious link to human well-being. Indeed, if we define as "risk" everything that could conceivably go wrong with the use of science and technology—a definition that some are evidently prepared to adopt—then it seems possible that we might arrive at a general understanding of limits to guide the moral aspects of scientific and technical practice. It is precisely because such an understanding is almost entirely lacking in modern thought that its possible development in the study of risk seems a hopeful prospect.

But this promise will be difficult to realize. The arena in which discussions of risk occur is by now highly politicized and contentious. In both specific questions like the safety of nuclear power, as well as more general ones that concern the proper methodology for studying risks at all, the stakes are high. Powerful social and economic interests are invested in attempts to answer the question "how safe is safe enough?" Expert witnesses on different sides of such issues are often best identified not by what they know, but rather by whom they represent. Indeed, the very introduction of risk as a common way of defining policy issues is itself far from a neutral development. At a time in which modern societies are beginning to respond to a wide range of complaints about possible damage to environment and public health caused by various industrial practices, the introduction of self-conscious risk assessment adds a distinctively conservative influence. By the term "conservative" here I mean simply a point of view that tends to favor the status quo. Although a good number of those who have become involved in risk assessment are not conservative in a political sense, it seems to me that under any foreseeable conditions in which this new art will be practiced, its primary effect will be to delay, complicate, and befuddle issues in a way that will sustain an industrial status quo relatively free of socially enforced limits. It is the character of this conservatism that I want to explore here.

Hazards and Consensus

As compared to other varieties of moral and political argument, risk assessment seeks a very narrow consensus. It asks us to evaluate circumstances in which there is some chance, perhaps a very remote chance, of harm from activities that are assumed to be socially beneficial in other respects. If one is able to recognize and care about the possibility of such harm, one is eligible to enter the discussion. No other grounds of agreement are necessary. It does not matter what one's views on other issues may be—what one thinks about arms control, deficit spending, or abortion, for instance. People who have drastically different views on, say, welfare payments for the poor and virtually every other social question may nevertheless find reason to act together on shared dangers to health and safety.

In this respect, risk assessment differs from attempts to provide a general evaluation of the conditions of modern life, for example, the attempts of liberalism, Marxism, or other broad-scale social theories. It sometimes happens, of course, that the question of risk is formulated as a problem to be discussed in the terms of a more comprehensive theory. A Marxist may reformulate the problem as a wrinkle in the analysis of relations of production; a utilitarian may wish to see it as a source of perplexity for the wish to achieve the greatest good for the greatest number. But such discussions are not prominent in claims and counter claims about specific hazards. We do not look to debates about DDT, PCBs, air pollution, nuclear power, and the like for a comprehensive grasp of the modern condition. The topic here concerns, as one observer has described it, "making industrialism safe for human life."[1]

But although the consensus that risk assessment seeks is a narrow one, it is potentially very strong. The fundamental issue here is fear: fear of injury, disease, death, and of the prospect of having to live in deteriorating surroundings. For that reason, arguments about impending dangers are often useful in attempts to unite people who have little in common other than shared fears. Contemporary environmentalism and consumerism have taken advantage of this opportunity to marshal the support for their causes. On a much different scale, the military has won legitimacy for its many multibillion-dollar projects by engaging the public's anxieties about ambiguously defined enemies and the peril of nuclear war.

To appreciate the remarkable strength of arguments predicated on fear, there is no better counsel than the great political psychologist of dread, Thomas Hobbes. It was Hobbes's fundamental insight that people who can agree on nothing else will nevertheless recognize that what is common to them is a morbid fear of physical harm from each other. Even the strongest person in the "state of nature" is vulnerable to attack in an unguarded moment. When people recognize the continuing terror that surrounds them, they will be ready to enter a compact establishing political society and its reliable system of authority and obligation. As the first and most brilliant modern writer on risk assessment, Hobbes gives us adequate reasons why people's fears should never be taken lightly.

Numerous examples in the history of modern industrial society lend support to the belief that overtly dangerous applications of new technology will not long be tolerated. Head-on collisions of passenger trains in New England in the mid-nineteenth century called attention to scheduling and communications problems for railroads of the time. The trains in question ran on single lines of track, a situation that made simultaneous two-way traffic a tricky matter even with occasional switchovers. Regulations passed by state legislatures enforced a widely appreciated need for remedies.[2] A similar set of events afflicted the early history of jet airline travel. An unforeseen weakness in the aluminum body of the British Comet caused several of the planes to fall from the sky. Study of the wreckage revealed the problem's source and the manufacturer redesigned its planes in response. Thus, the early stages of development of new machines, chemicals, techniques, and large-scale systems frequently involve a period of trial and error in which people are killed or injured. A commonly accepted norm is that obvious sources of harm must be eliminated through either private or public action; otherwise the very usefulness of the device will be called into question. And there are examples of technology—thalidomide is an obvious case—ultimately judged not a tool, but a menace.

In fact, the public response to dangerous effects of modern technological and industrial systems has often been a stimulus for reform when other sources were weak. Upton Sinclair's *The Jungle* (1906) aroused widespread protest about unhealthy conditions in the Chicago stockyards. Investigations by public officials led to new laws and remedies. Sinclair, a socialist, wrote dozens of books decrying the evil effects of capitalism in modern society. None of his later writings, however, was as successful in capturing the popular imagination or in stimulating change than his expose of the meat-packing industry.[3] From Sinclair's experience and that of other political activists more recently, it is clear that alarms about particular hazards will engage the public's imagination where more ambitious, general criticisms do not. Hence, the politics of hazards often becomes a strategic complement for or even an alternative to the politics of social justice.

The attempt to parlay the principle "salus populi suprema lex" into a full-blown political movement has become a familiar approach among contemporary activists. By calling attention to a possible danger, one hopes to attract support for a broader program of social criticism and reform. The alleged danger works as a symbol that may enable people to consider other social maladies, for example, the concentration of institutional power. Thus, the first sentence of Ivan Illich's *Medical Nemesis* proclaims, "The medical establishment has become a major threat to health."[4] Illich's method is to focus on "iatrogenic" (physician-caused) disease as a way to attract the reader's attention to his primary concern: the destructive social organization of modern medicine. Clearly, in Illich's view, this mode of organization—monolithic, bureaucratic, expert centered—would be pernicious wherever it occurred. The fact that it is also the source of certain identifiable health problems gives a compelling reason to explore the problem. Illich seizes upon this fact to instruct his audience about the oppressive social structures of modern life, not just in medicine, but throughout a whole range of institutionalized forms.

Similar approaches are now used with varying success by environmentalists, consumerists, and muckraking political journalists. Headline news about dangers to health and safety provides opportunities to discuss what these observers describe as even more fundamental issues; for example, the need to control the enormous power of large business corporations. Articles and editorials frequently published in *The Nation, The Progressive, Mother Jones,* or *In These Times* show this strategy of persuasion at work. And at one level it is perfectly sound. In a society strongly committed to capitalism as a way of life, it is difficult to address capitalist practices head-on. Thus, one begins by discussing urgent issues that do not appear to have any ideological valence at all.[5] Pointing to this strategy, I do not wish to dismiss any of the claims made by these political activists or writers. But it is fair to notice that this approach involves rolling out a Trojan horse now and again. If it were possible to tackle social injustice and the concentration of economic power more directly, we would, I suppose, do so.

Risk and Fortitude

But as political strategies sometimes do, this way of addressing social issues can backfire. The rise of risk assessment in the 1970s gave adequate notice that a strong backfire had begun. Questions that had previously been talked about in such terms as the "environmental crisis," "dangerous side effects," "health hazards," and the like were gradually redefined as questions of "risk."[6] The difference is of no small importance.

If we declare ourselves to be identifying, studying, and remedying hazards, our orientation to the problem is clear. Two assumptions, in particular, appear beyond serious question. First, we can assume that given adequate evidence, the hazards to health and safety are fairly easily known. Second, when hazards of this kind are revealed, agreement on what to do about them is readily available among all reasonable people. Thus, if we notice that a deep, open pit stands along a path where children walk to school, it seems wise to insist that the responsible party, be it a private person or public agency, either fill the pit or put a fence around it. Similarly, if we have good reason to believe that an industrial polluter is endangering our health or harming the quality of the land, air, or water around us, it seems reasonable to insist that the pollution cease or be strongly curtailed. Straightforward notions of this kind, it seems to me, lie at the base of a good many social movements concerned with environmental issues, consumer protection, and the control of modern technology. In other ways, of course, such movements are capable of adding elements of complication to policy discussion: for example, notions of complexity from ecological theory. Typically, however, these complications are ones that ultimately reinforce a basic viewpoint that sees dangers to human health, other species, and the environment as grave matters, fairly easy to understand and requiring urgent remedies.

If, on the other hand, we declare that we are interested in assessing risks, complications of a different sort immediately enter in. Our task now becomes that of studying, weighing, comparing, and judging circumstances about which no simple consensus is available. Both of the commonsense assumptions upon which the concern for hazards and dangers relies are abruptly suspended. All confidence in how much we know and what ought to be done about it vanishes in favor of an excruciatingly detailed inquiry with dozens (if not hundreds) of fascinating dimensions. A new set of challenges presents itself to the scientific and philosophical intellect. Action tends to be postponed indefinitely.

As one shifts the conception of an issue from that of hazard/danger/threat to that of risk, a number of changes tend to occur in the way one treats that issue. What otherwise might be seen as a fairly obvious link between cause and effect, e.g., air pollution and cancer, now becomes something fraught with uncertainty. What is the relative size of that risk, that chance of harm? And what is the magnitude of the harm when it does take place? What methods are suited to measuring and analyzing these matters in a suitably rigorous way? Because these are questions that involve scientific knowledge and its present limits, the risk assessor is constrained to acknowledge what are often highly uncertain findings of the best available research. For example, one must say in all honesty: "we don't know the relationship between this chemical and the harm it may possibly cause." In this way, the norms that regulate the acceptance or rejection of the findings of scientific research become, in effect, moral norms governing judgments about harm and responsibility. A very high premium is placed on not being wrong. Evidence that the experts disagree adds further perplexity and a need to be careful before drawing conclusions. The need to distinguish facts from values takes on paramount importance. Faced with uncertainty about what is known concerning a particular risk, prudence becomes not a matter of acting effectively to remedy a suspected source of injury, but of waiting for better research findings.[7]

An illustration of this cast of mind can be seen in a study done for the Environmental Protection Agency to determine whether or not there were indications that residents of the Love Canal area of New York, site of an abandoned chemical waste disposal dump, showed chromosome damage. The report written by Dante Picciano, a geneticist employed by the Biogenics Corporation of Houston, Texas, drew the following conclusions:

> It appears that the chemical exposure at Love Canal may be responsible for much of the apparent increase in the observed cytogenetic aberrations and that the residents are at an increased risk of neoplastic disease, of having spontaneous abortions and of having children with birth defects. However, in absence of a contemporary control population, prudence must be exerted in the interpretation of such results.[8]

Although the chemicals themselves may have been disposed of in reckless fashion, scientific studies on the consequences must be done with scrupulous care. Insofar as law and public policy heed the existing state of scientific knowledge about particular risks, the same variety of caution appears in those domains as well.

Frequently augmenting these uncertainties about cause and effect are the risk assessor's calculations on costs and benefits. To seek practical remedies for human-made risks to health, safety, or environmental quality typically requires an expenditure of public or private money. How much is it reasonable to spend in order to reduce a particular risk? Is the cost warranted as compared to the benefit received? Even if one is able to set aside troubling issues about equity and "who pays?", risk/cost/benefit calculations offer, in their very nature, additional reasons for being hesitant about proposing practical remedies at all. Because it's going to cost us, we must ponder the matter as a budget item. Our budgets, of course, include a wide range of expenditures for things we need, desire, or simply cannot avoid. Informed about the ways that the cost of reducing environmental risks is likely to affect consumer prices, taxes, industrial productivity, and the like, the desire to act decisively with respect to any particular risk has to be weighed against other economic priorities.[9]

A willingness to balance relative costs and benefits is present in the very adoption of the concept of risk to describe one's situation. In ordinary use, the word implies "chance of harm" from the standpoint of one who has weighed that harm against possible gain. What does one do with a risk? Sometimes one decides to take it. What, by comparison, does one do with a hazard? Usually one seeks to avoid it or eliminate it. The use of the concept of risk in business dealings, sports, and gambling reveals how closely it is linked to the sense of voluntary undertakings. An investor risks his capital in the hope of making a financial gain. A football team in a close game takes a risk when it decides to run on fourth down and a yard to go. A bettor at a Las Vegas blackjack table risks her money on the chance of a big payoff. In contrast to the concepts of danger, hazard, or peril, the notion of risk tends to imply that the chance of harm in question is accepted willingly in the expectation of gain. This connotation of the term makes the distinction outlined in some of the recent literature between voluntary and involuntary risks a largely misleading one. The word carries a certain baggage, a set of ready associations. The most important of these is the simple recognition that all of us take risks of one kind or another rather frequently.

Noticing that everyday life is filled with risky situations of various kinds, contemporary risk assessment has focused upon a set of psychological complications that further compound the difficulties offered by scientific uncertainty and the calculations of risk/cost/benefit analysis. Do people accurately estimate the risks they actually face? How well are they able to compare and evaluate such risks? And why do they decide to focus upon some risks rather than others? A good deal of interesting and valid psychological research has been devoted to answering such questions. By and large, these studies tend to show that people have a fairly fuzzy comprehension of the relative chance of harm involved in their everyday activities.[10] If one adds to such findings the statistical comparisons of injuries and fatalities suffered in different situations in modern life, then the question of why people become worried about certain kinds of risks and not others becomes genuinely puzzling.[11]

This puzzle is pregnant with rhetorical possibilities. It is often seized upon by writers who assert that people's confusion about risks discredits the claims of those who focus upon the chance of harm from some particular source. Why should a person who drives an automobile, a notorious cause of injury and death, be worried about nuclear power or the level of air pollution? Invidious comparisons of this kind are sometimes employed to show that people's fears about technological hazards are completely irrational. Hence, one leading proponent of this view argues:

> It is not surprising that people with psychological and social problems are unsettled by technological advance. The fears range from the dread of elevators in tall buildings to apprehension about "radiation" from smoke detectors. Invariably, these fears are evidence of displacing of inner anxiety that psychiatrists label as phobic.

The same writer explains that normal folk are able to overcome such phobias by reminding themselves of the incalculable good that modern technologies have brought to all of us. "People of sound mind accept the negligible risk and minor inconvenience that often go hand in hand with wondrous material benefits."[12]

Once one has concluded that reports about technological risks are based in phobia, the interesting task becomes that of explaining why people have such fears at all. Tackling this intellectual challenge, anthropologist Mary Douglas and political scientist Aaron Wildavsky have developed a style of analysis based on the assumption that complaints about risk are not to be taken at face value. In their view, all reports about environmental risks must be carefully interpreted to reveal the underlying social norms and institutional attachments of those making the complaints. Different kinds of institutions respond to risk in very much different ways. For example, entrepreneurs accept many kinds of economic risk without question. They embrace the invigorating uncertainties of the market, the institutional context that gives their activities meaning. In contrast, public interest organizations of the environmental movement, organizations that Douglas and Wildavsky describe as "sects," show, in their view, obsessive anxiety about technological risks; the discovery of these risks provides a source of personal commitment and social solidarity the sects so desperately need. Are there any environmental dangers in the world that all reasonable people, regardless of institutional attachment, ought to take seriously? Douglas and Wildavsky find that question impossible to answer. The fact that the scientists disagree requires us to be ever skeptical about any claims about particular risks. Instead, Douglas and Wildavsky offer the consolations of social scientific methodology to help us explain (and feel superior to) the strange behavior of our benighted contemporaries.[13]

Entering thickets of scientific uncertainty, wending our way through labyrinths of risk/cost/benefit analysis, balancing skillfully along the fact/

value gap, stopping to gaze upon the colorful befuddlement of mass psychology, we finally arrive at an unhappy destination: the realm of invidious comparison and social scorn. This drift in some scholarly writings on risk assessment finds its complement in the public statements and advocacy advertising of corporations in the oil, chemical, and electric power industries. In the late 1970s the debunking of claims about environmental hazards became a major part of corporate ideology. Closely connected to demands for deregulation and the relaxing of government measures to control air pollution, occupational safety and health, and the like, the risk theme in the pronouncements of industrial firms assumed major importance.

A typical advertisement from Mobil Oil's "Observations" series illustrates the way in which popularized risk psychology and rick/cost/benefit analysis can work in harmony. "Risky busness," the ad announces. "Lawn mowers . . . vacuum cleaners . . . bathtubs . . . stairs . . . all part of everyday life and all hazardous to your health. The Consumer Product Safety Commission says these household necessities caused almost a million accidents last year, yet most people accept the potential risks because of the proven benefits. . . . Risk, in other words, is part of life. Fool's goal. Nothing's safe all the time, yet there are still calls for a 'risk-free society.' " Although I have read large portions of the recent literature on energy, the environment, consumer protection, and the like, I cannot recall having seen even one instance of a demand for a risk-free society. The notion appears only as a straw man in advocacy ads like this one. Its text goes on to evoke a string of psychological associations linked to the experience of risk in economic enterprise. "Cold feet. What America does need are more companies willing to take business risks, especially on energy, where the risks are high. . . . We're gamblers. . . . Taking risks: it's the best way to keep America rolling . . . and growing."[14] Poker anyone?

There is, then, a deep-seated tendency in our culture to appreciate risk taking in economic activity as a badge of courage. Putting one's money, skill, and reputation on the line in a new venture identifies that person as someone of high moral character. On the other hand, people who have qualms about occasional side effects of economic wheeling and dealing can easily be portrayed as cowardly and weak-spirited, namby-pambies just not up to the rigors of the marketplace. Public policies that recognize such qualms can be dismissed as signs that the society lacks fortitude or that the citizenry has grown decadent. Thus, in addition to other difficulties that await those who try to introduce risk as a topic for serious political discussion, there is a strong willingness in our culture to embrace risk taking as one of the warrior virtues. Those who do not possess this virtue should, it would seem, please not stand in the way of those who do.

Avoiding "Risk"

By calling attention to these features in contemporary discussions about risk, I do not want to suggest, as some have done, that the whole field of study has somehow been corrupted by the influence of selfish economic interests.[15] Neither am I arguing that all or most conversations on this topic show a deliberate, regressive political intent. Indeed, many participants have entered the debate with the most noble of scientific, philosophical, and social goals. Much of the analytically solid writing now produced on this topic seeks to strengthen intellectual armaments used to defend those parts of society and the environment most likely to experience harm from a variety of technological side effects.[16] And certainly there are many fascinating issues under the rubric of risk assessment that are well worth pursuing. Of such clearheaded, magnanimous work, I can only join in wishing it flourish.

But from the point of view I've described here, the risk debate is one that certain kinds of social interests can expect to lose by the very act of entering. In our times, under most circumstances in which the matter is likely to come up, deliberations about risk are bound to have a strongly conservative drift. The conservatism to which I refer is one that upholds the status quo of production and consumption in our industrial, market-oriented society, a status quo supported by a long history of economic development in which countless new technological applications were introduced with scant regard to the possibility they might cause harm. Thus, decades of haphazard use of industrial chemicals provide a background of expectations for today's deliberations on the safety of such chemicals. Pollution of the air, land, and water are not the exception in much of twentieth-century America, but rather the norm. Because industrial practices acceptable in the past have become yardsticks for thinking about what will be acceptable now and in the future, attempts to achieve a cleaner, healthier environment face an uphill battle. The burden of proof rests upon those who seek to change long-existing patterns.

In this context, to define the subject of one's concerns as "risk" rather than some other issue skews the subsequent discussion in a particular direction. This choice makes it relatively easy to defend practices associated with high levels of industrial production; at the same time it makes it much more difficult for those who would like to place moral or political limits upon that production to make much headway. I am not saying that this is a consequence of the way risk assessment is used, although conservative uses of this sort of analysis are, as we have seen, easily enough concocted. What is more important to notice is that, in a society like ours, discussions centering on risk bear an inherent tendency that shapes the texture of such inquiries and their outcome as well. The root of this tendency lies, very simply, in the way the concept of risk is employed in everyday language. As I have noted above, employing this word to talk about any situation declares our willingness to compare expected gain with possible harm. We generally do not define a practice as a risk unless there is an anticipated advantage somehow associated with that practice. In contrast, this disposition to weigh and compare is not announced by concepts that might be employed as alternatives to risk—danger, peril, hazard, and threat. Such terms do not presuppose that the source of possible injury is also a source of benefits. From the outset, then, those who might wish to propose limits upon any particular industrial or technological application are placed at a disadvantage by selecting risk as the focus of their concerns. As they adopt risk assessment as a legitimate activity, they tacitly accept assumptions they might otherwise wish to deny (or at least puzzle over): that the object or practice that worries them must be judged in light of some good it brings and that they themselves are recipients of at least some portion of this good.

Once the basic stance and disposition associated with risk have defined the field of discourse, all the complications and invidious comparisons I have described above begin to enter in. Standards of scientific certainty are applied to the available data to show how little we know about the relationship of cause and effect as regards particular industrial practices and their broader consequences. Methods of risk/cost/benefit analysis fill out a detailed economic balance sheet useful in deciding how much risk is acceptable. Statistical analyses show the comparative probability of various kinds of unfortunate events, e.g., being injured in a skiing accident as compared to being injured by a nuclear power plant meltdown. Psychological studies reveal peculiarities in the ways people estimate and compare various kinds of risks. Models from social science instruct us about the relationship of institutional structures to particular objects of fear. A vast, intricately specialized division of intellectual labor spreads itself before us.

One path through this mass of issues is to take each one separately in its own right, seeking to determine which standards, methods, findings, and models are appropriate to making sound judgments about problems that involve public health, safety, and environmental quality. For example, one might question how reasonable it is to apply the very strict standards of certainty used in scientific research to questions that have a strong social or moral component. Must our judgments on possible harms and the origins of those harms have only a 5 percent chance of being wrong? Doesn't the use of that significance level mean that possibly dangerous practices are "innocent until proven guilty"?[17]

Similarly, one might reevaluate the role that cost/benefit analysis plays in the assessment of risks, pointing to the strengths and shortcomings of that method. How well are we able to measure the mix of costs and benefits involved in a given choice? What shall we do when faced with the inadequacy of our measurements? Are criteria of efficiency derived from economic theories sufficient to guide value choices in public policy? In controversies about the status of the intellectual tools used in decision making, such questions are hotly disputed.[18]

But for those who see issues of public health, safety, and environmental quality as fairly straightforward matters requiring urgent action, these exercises in methodological refinement are of dubious value. It is sensible to ask: why get stuck in such perplexities at all? Should we spend our time working to improve techniques of risk analysis and risk assessment? Or should we spend the same time working more directly to find better ways to secure a beautiful, healthy, well-provided world and to eliminate the spread of harmful residues of industrial life?

The experience of environmentalists and consumer advocates who enter the risk debate will resemble that of a greenhorn who visits Las Vegas and is enticed into a poker game in which the cards are stacked against him. Such players will be asked to wager things very precious to them with little prospect that the gamble will deliver favorable returns. To learn that the stacked deck comes as happenstance rather than by conscious design provides little solace; neither will it be especially comforting to discover that hard work and ingenuity might improve the odds somewhat. For some, it is simply not the right game to enter. There are some players at the table, however, who stand a much better chance. Proponents of relaxed government regulations on nuclear power, industrial pollution, occupational safety and health, environmental protection, and the like will find risk assessment, insofar as they are able to interest others in it, a very fruitful contest. Hence, Chauncey Starr, engineer and advocate of nuclear power, is well advised to take risk as the central theme in his repertoire of argument. But the likes of David Brower, Ralph Nader, and other advocates of consumer and environmental interests would do well to think twice before allowing the concept to play an important role in their positions on public issues.

Fortunately, many issues talked about as risks can be legitimately described in other ways. Confronted with any cases of past, present, or obvious future harm, it is possible to discuss that harm directly without pretending that you are playing craps. A toxic waste disposal site placed in your neighborhood need not be defined as a risk; it might appropriately be defined as a problem of toxic waste. Air polluted by automobiles and industrial smokestacks need not be defined as a risk; it might still be called by the old-fashioned name, "pollution." New Englanders who find acid rain falling on them are under no obligation to begin analyzing the risks of acid rain; they might retain some Yankee stubborness and confound the experts by talking about "that destructive acid rain" and what's to be done about it. A treasured natural environment endangered by industrial activity need not be regarded as something at risk; one might regard it more positively as an entity that ought to be preserved in its own right.

About all these matters there are rich, detailed forms of discourse that can help inform our judgments and provide structure for public decisions. A range of theoretical perspectives on environmental protection, public health, and social justice can be drawn upon to clarify the choices that matter. My suggestion is that before risk is selected as a focus in any area of policy discussion, other available ways of defining the question be thoroughly investigated. For example, are health and safety hazards that blue-collar workers encounter on the job properly seen as a matter of risk to be analyzed independently of directly related economic and social conditions? Or is it more accurate to consider ways in which these hazards reflect a more general set of social relationships and inequalities characteristic of the free enterprise system? One's initial definition of the problem helps shape subsequent inquiries into its features. If one identified the issue of worker health and safety as a question of social justice, there would be less need to do all of the weighing of probabilities, comparing of individual psychological responses, and other delicate tasks that risk assessment involves. It might still be interesting to do research on levels of air pollution in executive offices as compared to those on factory assembly lines. Of course, one always wants to have the best scientific information on questions like these. But, in all likelihood, such studies would reveal little new or surprising. It is common knowledge that our society distributes wealth, income, knowledge, and social opportunities unequally. To establish that it also distributes workplace hazards inequitably merely amplifies the problem. Those concerned with questions of social justice would do well to stick to those questions and not look to risk analysis to shed much light.

My point is, then, that a number of important social and political issues are badly misdefined by identifying them as matters of risk. Whenever possible, such misdefinitions ought to be resisted along with the methodological quagmires they bring in train. This is not to say that there are no issues of broad-scale social policy in which the concept of risk is legitimately applied. There are some applications of modern science and technology in which the uncertainty that surrounds suspect practices and their possible effects is so great that risk is an entirely suitable name for what is problematic. Recent worries about possible mishaps from the use of recombinant DNA techniques in scientific research and industrial applications seem to me a case in which the term was accurately applied. But there is, in our time, a willingness to cluster an astonishingly large range of health, safety, and environmental problems under this one rubric. Thus, Douglas and Wildavsky, among other observers, rush to the conclusion that all environmentalists ever worry about are risks. Misconceptions of that sort are ones we ought to avoid.

If the conservative voice in risk assessment were one that counseled us to cherish and preserve those parts of our culture, society, and environment that are genuinely worth saving, then I would, as a devoted conservative, view this work with much hope. Unfortunately, the style of conservatism we all too often find here is in the "damn the torpedoes" tradition of economic and technological thinking. Rather than lead us to devise better ways of expressing caution and care, such thinking often justifies a renewed recklessness.

Two kinds of issues in particular strike me as territory that ought to be rescued from this tendency. First are our understandings of cases of actual harm—cancer, birth defects, other illnesses, deaths, damaged environments, and the like—obviously connected to profit-making industrial practices, yet sometimes treated as if their reality were merely probabilistic. We may visit the hospitals and gravesites, if we need to. We may wander through industrial wastelands and breathe deeply. But let us not pretend our troubles hinge on something like a gentlemanly roll of the dice, or that other people's sickness and death can be deemed acceptable from some august, supposedly neutral standpoint. That only adds insult to injury.

Finally, there are times in which an issue that urgently needs to be addressed is poorly described (or not described at all) by subsuming it under the category of risk. During the past decade or so a number of important items on the public agenda have been affected in this manner. Many of the interesting social problems linked to the development of nuclear power, for instance, have nothing to do with risk as such. But to a large extent the debate over risk and safety informed the range of public discussions on nuclear power to the exclusion of everything else.

A similar warping of the public's attention may well happen on another crucial question now before us. A technology recognized above as an example of true risk in one context, recombinant DNA research and development, will be badly misconstrued if it is seen as nothing more than a risk question in another emerging context: public policy on genetic engineering. It is one thing to think about the prospect that a lethal bug might escape from the laboratory, quite another to ponder what it means to assume direct control of the evolution of the human species. Possibilities made available by the new biotechnologies are profound ones. But they are not questions of risk. It may happen, nevertheless, that because risk was the focus of discussion for the first decade of thinking about the ethical dimensions of recombinant DNA research and development, it will continue to shape discussions on this topic for years to come.[19] We could see, for example, the application of moral rules of the following kind: unless the development of a new genetic configuration can be shown to involve substantial quantifiable risks, the development will be sanctioned for speedy implementation. If that happens, the shortcomings of today's discussions about risk will return to us with a vengeance.

Afterword

"Didn't the fox *never* catch the rabbit, Uncle Remus," the little boy asked the next evening.

"He came mighty nigh it, honey, sho's you born."

Thus begins "The Wonderful Tar-Baby Story," an American classic told by Joel Chandler Harris. In the tale, Brer Fox hatches a scheme to fool his archrival, elusive Brer Rabbit. Using a mixture of tar and turpentine, he fashions a little figure, places it by the side of the road, and hides in the bushes to watch. Sure enough, before long Brer Rabbit comes pacing down the road and says good morning to the dark stranger. But "The Tar-Baby ain't sayin' nothin' en Brer Fox, he lay low."

Brer Rabbit takes the silence as an insult. After making repeated demands for a response, he hauls off and punches the Tar-Baby and gets his paw stuck. He tries another paw with the same result, then both hind feet and finally his head.

" 'Howdy, Brer Rabbit,' sez Brer Fox sezee. 'You look sorter stuck up dis mawning,' sezee, en den he rolled on de groun', en laughed twel he couldn't laugh no mo.' "

"Did the Fox eat the Rabbit?" asked the little boy who had been listening to the story with rapt attention.

"Dat's all de fur de tale goes," replied the old man.[20]

Notes

This chapter was prepared with the support of a grant from the National Science Foundation and National Endowment for the Humanities program in Ethics and Values in Science and Technology, OSS-8018089. Its views are those of the author and do not necessarily reflect those of NSF or NEH. I want to thank Dudley Burton, Mary Gibson, Tom Jorling, Frank Laird, James O'Connor, and Tim Stroshane for their comments and suggestions on an earlier draft.

1. Samuel S. Epstein, Lester O. Brown, and Carl Pope, *Hazardous Waste in America* (San Francisco: Sierra Club Books, 1982), p. 6.
2. Stephen Salsbury, *The State, the Investor, and the Railroad: Boston & Albany, 1825–1867* (Cambridge, Mass.: Harvard University Press, 1967), pp. 182–87.
3. Upton Sinclair, *The Jungle* (New York: Doubleday, 1906). See also Leon A. Harris, *Upton Sinclair, American Rebel* (New York: Crowell, 1975).

4. Ivan Illich, *Medical Nemesis: The Expropriation of Health* (New York: Pantheon Books, 1976), p. 3. "Increasing and irreparable damage," Illich writes, "accompanies present industrial expansion in all sectors. In medicine this damage appears as iatrogenesis. Iatrogenesis is clinical when pain, sickness, and death result from medical care; it is social when health policies reinforce an industrial organization that generates ill-health. . . ." pp. 270-71.

5. See, for example, Mark Green and Robert Massie Jr., eds., *The Big Business Reader* (New York: The Pilgrim Press, 1980), especially "And Filthy Flows the Calumet" by Edward Greer, pp. 169-79; "Love Canal and the Poisoning of America" by Michael H. Brown, pp. 189-207; "The Dumping of Hazardous Products on Foreign Markets" by Mark Dowie and *Mother Jones*, pp. 430-43.

6. Two important writings that helped place the concept of risk at the center of discussions about technology, environment, and society are: Chauncey Starr, "Social Benefit versus Technological Risk," *Science* 165 (September 19, 1969): 1232-38; William Lowrance, *Of Acceptable Risk* (Los Altos, Calif.: William Kaufmann, Inc., 1976).

7. Steven D. Jellinek laments this state of affairs in "On the Inevitability of Being Wrong," *Technology Review*, August/September 1980, pp. 8-9.

8. Dante Picciano, "Pilot Cytogenetic Study of Love Canal, New York," prepared by the Biogenics Corporation for the Environmental Protection Agency, May 14, 1980; quoted in Epstein, et al., *Hazardous Waste in America*, pp. 113-14.

9. See Edmund A. C. Crouch and Richard Wilson, *Risk/Benefit Analysis* (Cambridge, Mass.: Ballinger Publishing Co., 1982).

10. See Baruch Fischhoff et al., *Acceptable Risk* (New York: Cambridge University Press, 1981).

11. Chauncey Starr, Richard Rudman, and Chris Whipple, "Philosophical Basis for Risk Analysis," *Annual Review of Energy* 1 (1976): 629-62.

12. Samuel C. Florman, "Technophobia in Modern Times," *Science '82*, April 1982, p. 14.

13. Mary Douglas and Aaron Wildavsky, *Risk and Culture: The Selection of Technical and Environmental Dangers* (Berkeley: University of California Press, 1982). See my review, "Pollution as Delusion," *New York Times Book Review*, August 8, 1982, pp. 8 and 18.

14. From "Observations," an advertisement of the Mobil Oil Corporation, *Parade Magazine*, December 12, 1982, p. 29.

15. David Noble argues this position forcefully in "The Chemistry of Risk: Synthesizing the Corporate Ideology of the 1980s," *Seven Days* 3, no. 7 (June 5, 1979): 23-26, 34. A similar view is offered by Mark Green and Norman Waitzman, "Cost, Benefit and Class," *Working Papers for a New Society*, May/June 1980, pp. 39-51.

16. Intentions of this sort are evident in the research on risk assessment of The Center for Technology, Environment, and Development at Clark University. See Patrick Derr, Robert Goble, Roger E. Kasperson, and Robert W. Kates, "Worker/Public Protection: The Double Standard," *Environment* 23, no. 7 (September 1981): 6-15, 31-36; Julie Graham and Don Shakow, "Risk and Reward: Hazard Pay for Workers," *Environment* 23, no. 8 (October 1981): 14-20, 44-45.

17. Talbot Page discusses the issues involved in seeking a balance between "false positives and false negatives" in judgments about risks in "A Generic View of Toxic Chemicals and Similar Risks," *Ecology Law Quarterly* 7, no. 2 (1978): 207-44.

18. Among the helpful criticisms of cost/benefit analysis are: Mark Sagoff, "Economic Theory and Environmental Law," *Michigan Law Review* 79, no. 7 (June 1981): 1393-419; and Steve H. Hanke, "On the Feasibility of Benefit-Cost Analysis," *Public Policy* 29, no. 2 (Spring 1981): 147-57.

19. Sheldon Krimsky, *Genetic Alchemy* (Cambridge, Mass.: The MIT Press, 1982).

20. Joel Chandler Harris, *The Complete Tales of Uncle Remus*, compiled by Richard Chase (Boston: Houghton Mifflin, 1955), pp. 6-8.

Index

Abrasive blasting, 211
Acid rain, 282–83
  current damages from, 55–56
  emission reduction and, 59–60
  future damages from, 56–57
  impact of, 52, 55–56, 63–65
  impact of reducing levels of, 68–71
  pollutants causing formation of, 57–58
  research on, 60–61
  water resource damages from, 56–57, 61–63
Acute effects
  of airborne chemical contaminants, 34, 35, 45
  of passive smoking, 5–6
Administrative controls, 34
Advertising
  for cigarettes, 10–11, 12, 14
  environmental hazards and, 279
Allergic alveolities, 35
Allergy, to tobacco smoke, 6, 15
Aluminum, 62, 65, 67
Ambient air standards
  emissions controls to achieve, 253–54
  flexibiliity of, 251–53
  workplace, 243–44, 251, 258
Ambient monitoring, 33, 187
Ambient pollution rights, 243, 258
American Cancer Society, 8–9
American Conference of Governmental Industrial Hygienists (ACGIH), 45, 47–48
American Lung Association, 12
American Petroleum Institute v. *Costle*, 225, 230
Anesthetic gases, 37, 49
Angina, 6, 155
Angiosarcoma, 37
Aplastic anemia, 37

Appeal rights, 212
Arkansas, 71
Aromatic amines, 37
Arsenic, 36, 38, 45, 47, 215
Arsine gas, 37
ASARCO, 215
Asbestos, 26, 35, 36, 39, 42, 45, 47, 49, 187, 192
  legal cases concerning, 204–5
Asbestosis, 47, 205
ASH (Action on Smoking and Health) 9, 11–12, 14
Asthma
  occupational, 35
  tobacco smoke and, 6
Authority, relations of constraint and, 146
Automobile Calendar, 206–7
Autonomy, 141–68
  conditions for, 150–54
  defined, 141
  heteronomy versus, 144–45
  internal/external aspects of, 147–50
  Lockean view of natural rights and, 91, 115
  outdoor air pollution and, 162–66
  passive smoking and, 154–57
  as a social product, 145–47
  workplace air pollution and, 157–62

Barad, C. B., 14–15
Baram, Michael, 161, 186–97
Benzene, 34, 37, 39, 46, 49
Benzidine, 37
Best interests, in informed consent, 170
Biogenics Corporation, 276
Biological monitoring, 33, 187, 192, 228
Bladder function, 37
Blood formation, 37

286 Index

Boundaries
 development of, 146–47
 internal/external distinctions and, 147–50
 natural rights concept and, 90, 92–94, 99, 101–3, 105, 115
Bronchitis, 35, 42
Brower, David, 282
Bureau of Labor Statistics, 202
Byssinosis, 35, 45, 213

Cadmium, 37, 38
Calcium, 65
California, 215, 226
Cancer, 202, 205
 acceptable levels of risk of, 23
 from active smoking, 5, 19
 from airborne contaminants in the workplace, 35–37, 38, 42, 43, 47, 186
 concept of common property and, 97
 from passive smoking, 9–10, 19, 20, 21, 23
Capitalism, 274
Carbon dioxide, 35
Carbon disulfide, 37, 38
Carbon monoxide, 37, 38, 39
 standards for airborne, 224
Carbon tetrachloride, 36, 37
Cardiac function
 airborne workplace contaminants and, 37
 angina and, 6, 155
 transported air pollutants and, 67
Cardiovascular disease, 5, 15
Carter, Jimmy, 206–7, 211, 224
Charge systems, 257–58
 ethical considerations in, 254–56
Children, workplace contaminants and, 38, 39
Chlorine ammonia, 35
Chloroform, 37
Chronic effects
 of airborne chemical contaminants, 34, 35–38, 42, 43, 45–49, 186–87
 of passive smoking, 6–10
Cirrhosis of the liver, 42
Civil Aeronautics Board, 11
Clean Air Act, 7, 19, 20, 52, 53, 58, 60, 246, 256
 goals of, 226
 National Ambient Air Quality Standards of, 7, 222–32
Clean Water Act, 250, 255
Coal dust, 35
Coal industry, 32, 52, 69–70
Coal Mine Safety and Health Act, 32
Coke oven emissions, 35–36, 45, 47
Coleman, Jules, 137–38
Collective bargaining, 203, 212–18
 joint labor-management committees and, 215–17, 218
 local enforcement mechanisms and, 212–15
 problems of, 217–18
Commitment, autonomy and, 148–50
Common law
 consent in, 188, 190–93
 profit maximization versus, 204
Common property, 95–100, 117
Compensation
 consent and, 112–17, 157–58
 for exposure to risk, 113–17, 157–58, 205
 no-fault, for asbestos claims, 205
 as remedy for harms incurred, 188
 workers', 44, 190, 192, 193–94, 204, 205, 244, 256
Connecticut, 71, 246
Consensus standards, 210–12
Consent, 238
 in accepting risk, 20–23
 common law and, 188, 190–93
 compensation and, 112–17, 157–58
 as component of good life, 76, 77–82, 83–87, 142
 individual rights and, 76–77, 80–82, 83–85, 142
 instrumental value of, 75–76, 77–80, 82–83, 85–87, 141–42
 outdoor air pollution and, 162–66
 passive smoking and, 154–57
 role of autonomy in, 141–54
 tacit, and natural rights, 93–94, 99–100, 108–11
 for use of common property, 98–100, 117
 workplace air pollution and, 157–62, 187–90
 see also Informed consent
Consumer Product Safety Act, 12
Consumer Product Safety Commission, 279
Contracts
 role of consent in, 187–88, 190
 workplace air pollution and, 47–49
 see also Unions
Controlled Substance Act, 12
Copper, 42, 67, 215
Copper salts, 36
Cost-benefit analysis
 in air quality standards, 222, 227–29, 230
 in control of airborne pollution, 235–36, 237, 262–63
 in imposition of risk without consent, 77–80, 87
 for risk reduction, 276–79, 281–82
Costle, Douglas, 225, 230, 246
Cotton dust, 35, 45, 46, 213
Court system

common law and, 188, 190–93, 204
tort law and, *see* Tort law
for workplace air pollution, 15–19, 44–47, 203–6
Crops
acid rain impact on, 55–56, 63–64
ozone impact on, 63, 64
Cytotoxic drugs, 37

DBCP, 37
Decision making
cognitive barriers to, 172–80
informed consent in, 170–71, 182–83
Department of Health and Human Services
Social Security Administration, 15–16
see also National Institute for Occupational Safety and Health
Department of Labor
Bureau of Labor Statistics, 202
see also Occupational Safety and Health Administration
Derr, Patrick, 188–89
Determinism
personal autonomy in, 141–44, 148
see also Decision making
Dioxin, 37
Disability benefits
from effects of passive smoking, 16–17
from occupational disease, 44
problems in collecting, 217
see also Workers' compensation
Discharge permits
marketable, 240–44, 257–58
nonmarketable, 252
Disclosure
chemical labeling standards in, 42, 43, 161–62, 195, 211
informed consent and, 183, 189, 192–93
of occupational contaminants to workers, 42–43
Disease rates, occupational exposure and, 39–44
DNA, 283, 284
Doll, Richard, 8
Dosimeters, 248
Douglas, Mary, 278, 284
Dworkin, Ronald, 137–38

Eads, George, 222–32
Emissions control, 59–60
ambient pollution rights for, 243, 258
automobile, 229
impact of increasing, 68–71
need for mathematical models in, 240
permits for, 240–42, 243–44, 252, 257–58
political and ethical appeal of, 253–57
problems of monitoring in, 245–49

for smokers, 244–45, 249, 258
uncertainties concerning, 60–61
Emphysema, 35
Employment, emissions controls and, 70
Engineering controls, 33–34, 213–14, 215
Environmental Protection Agency (EPA), 19
cigarette smoking pollution and, 19
Clean Air Act standards and, 224, 226–27, 228–29, 230–31, 246
Love Canal study, 276
Occupational Safety and Health Administration versus, 206
offset system of, for ambient air standards, 252
outdoor air pollution and, 58
Equal Employment Opportunity Commission, 49
Ethical considerations, 253–55
Ethylene oxide, 37, 49
Europe, 210–11

Fair Packaging and Labeling Act, 12
Fear
of being harmed, 105–7, 278
phobias and, 278
risk assessment and, 272–74
Federal Hazardous Substances Act, 12
Federal Rehabilitation Act of 1973, 15
FENSR (Federal Employees for Nonsmokers' Rights), 14
Fischhoff, Baruch, 162, 169–85
Florida, 68, 71
Fluorocarbon, 37
Food, Drug, and Cosmetics Act, 12
Forests
acid rain damage to, 57, 67–65
ozone damage to, 56, 64, 65
Friedman, Robert M., 52–72

Garfinkel, Lawrence, 8–9
GASP (Group Against Smokers' Pollution), 11–12, 14
Gibson, Mary, 141–68
Goodson, R. Eugene, 227
Gordan, Adele, 18
Gray, C. Boyden, 211–12
Great Britain, 211
Group Health Cooperative, 26

Hamilton, Alice, 31
Harris, Joel Chandler, 285
Health and safety committees, 215–17, 218
Health-care professionals, occupational disease and, 43, 44
Heteronomy
autonomy versus, 144–45
defined, 144
relations of constraint and, 146

Hirayama, Takeshi, 7–9, 20
Hobbes, Thomas, 273
Hume, David, 93
Hydrocarbons, 38, 53
Hypertension, 37

Illich, Ivan, 274
Illinois, 68, 70
Incentive systems, 238–39, 249–51
Indiana, 68, 69, 70
Industrial policy, 203, 206–9
 activist form of, 207–9
 regulatory objectives in, 214–15
Informed consent, 169–85
 aids to, 180–82
 characteristics of, 159
 cognitive barriers to, 172–80
 concept of, 170–71
 disclosure requirements and, 183, 189, 192–93
 information needed in, 189
 as matter of optimal decision making, 182–84
 see also Consent
Insurance
 no-fault, for asbestos claims, 205
 see also Workers' compensation
Intent, 111–12
Interagency Regulatory Liaison Group, 207
Interviews, informed consent in, 174–76

Jungle, The (Sinclair), 273–74

Kant, Immanuel, 106–7, 116, 117, 144
Kasperson, Jeanne X.,, 164
Kasperson, Roger, E., 164
Kentucky, 68, 69, 70
Kepone, 37
Kidney function, 36, 37

Latency period, 43, 186
Lead, 31, 36, 37, 38, 39, 45, 48, 49, 67, 215
Leukemia, 37
Liability system, 244, 248–49, 256
 see also Tort law
Liver function, 36–37, 42
Locke, John, 89–123, 237
Love Canal, 276

MacCarthy, Mark, 161, 201–21
Magnesium, 42, 65
Maine, 71
Manganese, 38
Mantel, Nathan, 8
Manville Corporation, 205
Massachusetts, 42–43, 48
Materials damage
 from air pollution, 65–66
 see also Acid rain; Ozone
Medical Nemisis (Illich), 274
Mercury, 37, 38, 67
Mesothelioma, 35
Metal-fume fever, 42
Methane, 35
Methyl butyl ketone, 38
Michigan, 71
Minnesota, 14
Missouri, 68
Mobil Oil, 279
Models
 in emissions control studies, 240
 of impact of airborne pollutants, 58, 60
Monitoring
 ambient, 33, 187
 biological, 33, 187, 192, 228
 problems of, 245–49
 self-, 246
Montana, 215
Moore, G. E., 128–30, 131, 132, 135
Mortality
 from active smoking, 5, 20
 death certificate information on, 44
 from occupational disease, 41–42, 202
 from passive smoking, 9, 10, 20, 21, 23
 from transported air pollutants, 67

Nader, Ralph, 282
Naphthylamine, 37
National Ambient Air Quality Standards (NAAQS), 7, 222–32
 cost-benefit analysis and, 222, 227–29
 enshrinement of, as goals, 226–27
 establishment of, 223, 224
 importance of statutory structure for, 229–31
 setting versus attaining, 225
 state implementation plans for, 224–25
 statutory requirements for, 224
National Institute for Occupational Safety and Health (NIOSH), 32, 202
 evaluation of standards and exposure levels by, 45
 surveys of occupational chemical exposure by, 39–41
National Labor Relations Act, 48–49
National Safety Council, 202
Natural rights, 89–123
 concept of common property and, 95–100
 dispositional harms and, 94–95
 Lockean view of, 90–92, 158
 personal boundaries and, 90, 92–94, 99, 101–3, 105, 115
 revised view of Lockean, 104–17
 risk and responsibility in exercising, 101–3
 see also Rights
Naval Research Laboratory, 7

Nervous system function, 38
New Jersey Bell Telephone, 17–18
N-hexane, 39
Nitrogen oxides, 35, 53
  see also Acid rain

Occupational Safety and Health Act (OSHA), 32, 43, 46, 48, 161, 194, 201, 202, 214
Occupational Safety and Health Act (OSH Act), 32, 43, 46, 48, 161, 194, 201, 202, 214
Occupational Safety and Health Administration (OSHA), 32, 194–95, 201–21
  chemical labeling standards proposed by, 42, 43, 161–62, 195, 211
  collective bargaining versus, 203, 212–18
  consensus standards versus policies of, 203, 208–9, 210–12
  court system versus, 203–6
  criticisms of, 201–3
  establishment of chemical exposure standards by, 45–49
  industrial policy versus, 203, 206–9
  log of recordable injuries and illnesses of, 43
  regulation of hazardous substances by, 42, 43, 44
Office of Technology Assessment (OTA)
  air pollutant damage to plants and, 64
  air pollutant damage to water resources and, 62–63
  analysis of mortality rates due to air pollutants, 67
  estimate of cost of emissions controls, 68–69, 71
  study of airborne pollutants of, 53
Office on Smoking and Health, 12
Offset system, 252
Ohio, 68, 70
Oil, Chemical, and Atomic Workers International Union, 215–16
Optimality in informed consent, 173, 177, 180–84
Outdoor air pollution, 7, 52–72
  action needed to control, 234–35
  concept of common property and, 96–97
  determining action taken on, 236–38
  ease of monitoring and enforcement of policies on, 238, 245–49
  efficiency of implementing program for, 238, 240–45
  flexibility of programs to control, 238–39, 251–53
  incentives to reduce, 238–39, 249–51
  inducing action on, 235–36
  information intensity of program on, 238, 240–45

  personal autonomy and consent in, 162–66
  personal boundaries and, 92–93
  political and ethical appeal of programs on, 239, 253–57
  risk and uncertainty in, 60–61
  risks of damage and risks of control in, 61–71
  scientific uncertainties of, 55–61
  transported pollutants in, 53–54
  see also Clean Air Act; Emission control; National Ambient Air Quality Standards
Ozone, 35
  current damages from, 55–56
  future damages from, 56–57
  impact of, 52, 55–57, 63, 64, 65
  pollutants causing formation of, 57–58
  standard for, in air, 224, 225, 227, 230

Parodi, Irene, 15–16
Paternalism
  in decision making, 183
  Lockean view of natural rights and, 91, 101, 114
PCBs (polychlorinated biphenyls), 36
Penalties, noncompliance, 246, 247
Pennsylvania, 68, 70, 71
Permits
  marketable, 240–44, 257–58
  nonmarketable, 252
Pesticides, 36, 37, 38, 39
Peters, R. S., 144
Petrochemicals, 45
Phobias, 278
Phosphorus, 36
Piaget, Jean, 145–47, 150
Picciano, Dante, 276
Pneumoconioses, 35
Pollution-control expenditures
  public financing of, 209
  reponsibility for, 208–9
Posner, Richard, 137
Pott, Percival, 31
Primary pollutants, 53
Prosser, William, 191–92
Prostate gland, 38
Protective equipment, 214–15
  respirators, 18, 34, 214
Psychiatric problems
  from fear of being harmed, 105–7, 278
  phobia as, 278
  work-related, 38, 43
Pulmonary disease
  from airborne chemical contaminants, 35–36, 42, 43, 47, 186
  from smoking, 5
  from transported air pollutants, 67

290 Index

Punnett, Laura, 31–51, 161

Questionnaires, informed consent in, 174–76

Railton, Peter, 87–123, 158, 162
Ramazzini, Bernardino, 31
Raven Systems and Research, Inc., 18
Rawls, John, 188
Reagan, Ronald, 211, 224
Reciprocal risk, 108
Repace, James L., 3–30
Reproductive function, 36, 37–38, 48, 49, 186
Respirators
  to prevent passive smoking, 18
  for workplace airborne chemicals, 34, 214
Richey, Charles R., 16
Rights
  to appeal regulatory standards, 212
  imposition of risks to others and, 124–40
  role of consent in, 76–77, 80–82, 83–85, 142
  workplace hazards and, 159–61
  see also Natural rights
Risk
  acceptable levels of, 23, 108–11
  assessment of, *see* Risk assessment
  compensation for exposure to, 113–17, 157–58, 205
  concept of, 54–55
  consent in acceptance of, 20–23
  cost-benefit analysis of, *see* Cost-benefit analysis
  decision concerning tolerability of, 164–65
  in decision making, 171
  in establishing occupational contaminant exposure levels, 45–47, 238
  of further emissions reductions, 68–71
  imposition of impure, 124, 125–26
  imposition of pure, 124, 126–40
  natural rights and, 95, 99–100, 101–3
  of no action concerning airborne pollutants, 61–67
  quantification of, 55
  reciprocal, 108
  role of consent in legitimation of, 75–88
Risk assessment, 271–86
  avoiding risk and, 280–85
  fortitude and, 275–79
  hazards and consensus in, 272–74
Roosevelt, Eleanor, 3
Russell, Clifford S., 233–68

Sampling techniques, 247–48
Scheffler, Samuel, 75–88, 141–42, 166
Secondary pollutants, 53–54
Self-defense, 104–8
Self-monitoring of polluters, 246
Selikoff, Irving, 204–5

Shimp, Donna, 16–17
Sidestream smoke. *See* Smoking, passive
Silica, 35, 211
Silicosis, 221
Sinclair, Upton, 273–74
Smith, Paul, 18, 19
Smoking, active
  cancer from, 5, 19
  growth of, 3–5
  imposition of risks to others through, 126
  incentives to limit, 250, 251
  monitoring of, 244–45, 249, 258
  mortality from, 5, 20
  occupational diseases and, 42, 43
  state legislation on, 14–15
Smoking, passive, 3–30, 241
  cancer from, 9–10, 19, 20, 21, 23
  effects of, on non-smokers, 5–10
  emissions controls for, 244–45, 249, 258
  flexibility of ambient standards for, 251–52
  growth of tobacco use and, 3–5
  mortality from, 9, 10, 20, 21, 23
  personal autonomy and consent in, 154–57
  political and ethical considerations for, 256–57
  problems of reducing risk of, 10–13
  strategies for reducing, 13–26
  in work environment, *see* Workplace air pollution
Socialism, 274
Social Security Administration, 15–16
Solvents, 36, 38
Starr, Chauncey, 282
State Implementation Plans (SIPs), 224–25
States
  Clean Air Act and, 58, 224–25
  emission control standards of, 253–54
  impact of emission controls on, 68–71
  legislation of, restricting smoking, 14–15
  workplace regulation by, 32, 42–43, 48
  see also Workers' compensation; specific states
Steroids, 37
Sulfur dioxide, 35, 47, 52, 53, 54
  reducing emissions of, 59–60, 68–71
  see also Acid rain
Survey research, informed consent and, 174–76
Sweden, 210–11, 217
Sweeney, Phillip J., 18

Task Force on Regulatory Relief, 211
Taxes, tobacco, 13–14, 114, 249, 256–57
Tennesee, 68, 70
Teratogens, 37–38
Thomson, Judith Jarvis, 134–40, 162
Tin compounds, 37

TNT, 37
Tobacco Institute, 8, 10–12
Tobacco Observer, The, 10–11
Toluene, 34
Tort law, 188, 190–93
  liability system and, 244, 248–49, 256
  Occupational Safety and Health Administration regulation versus, 195
  workers' compensation system versus, 193–94
Toxic psychosis, 38
Toxic Substance Control Act, 12
Toxic waste disposal sites, 276, 282
Trichloroethylene, 37
Trichopoulos, D., 8

Uncertainty
  concept of, 54
  quantification of, 5
  about transported air pollutants, 55–61
  see also Risk
Unions
  collective bargaining by, 203, 212–18
  role of, in occupational health hazards, 48, 49, 186–87
United Auto Workers, 216
U. S. Defense Department, nonsmoker protest in, 16–17
U. S. Surgeon General Advisory Committee on Smoking and Health, 5
  passive smoking and, 10
  Report on Smoking and Health of 1964, 14
United Steel Workers, 215, 216
Uranium, 36
Utilities, emissions controls, and, 68–69, 71

Variances, regulatory, 213
Ventilation systems, 33–34, 213–14, 215
Vermont, 71, 254
Veterans' Administration, nonsmoker protest in, 16
Vickers, Lanny, 15
Vinyl chloride, 37, 38, 45
Virginia, 70, 71
Visibility, air pollutant reductions in, 66
Voluntary compliance, 245

Voluntary consent, 191–92, 195
Voorhees, Donald, 15

Wald, Nicholas J., 9
Water resources
  acid rain damages to, 56–57, 61–63
  health effects of contamination of, 67
Western Electric, nonsmoker protest and, 18, 19
West Virginia, 68, 69, 70
Wildavsky, Aaron, 278, 284
Winner, Langdon, 271–86
Women
  smoking by, 3, 5
  workplace contaminants and, 38, 39, 48, 49
Workers' compensation, 190, 192, 193–94, 256
  lack of preventive incentive and, 205
  occupational diseases and, 44, 204
  problems of, 244
Workplace air pollution, 31–51, 186–97, 241
  ambient standards in, 243–44, 251, 258
  contractual remedies for, 47–49
  control measures for, 33–34, 187
  disease rates in, 39–44
  flexibility of ambient standards for, 251
  health hazards in, 34–38, 186–87
  issue of consent in, 157–62, 187–90
  legal remedies for, 15–19, 44–47, 203–6
  legislation concerning, 14–15, 19–20
  occupational exposure in, 39, 45–49
  Occupational Safety and Health Administration and, 194–95
  personal autonomy and consent in, 157–62, 187–90
  political and ethical considerations for, 256
  tobacco smoke in, 7, 12, 14–19, 23–26
  types of contaminants in, 32–33, 186
  workers' compensation claims and, 44, 190, 192, 193–94, 204, 205, 244, 256
  see also Occupational Safety and Health Administration

Zinc, 38, 42

Notes on Contributors

Michael Baram is adjunct professor of law at the Center for Law and Health Sciences of Boston University School of Law and professor at Boston University School of Public Health. He has written widely on federal regulation of environment and health and on state and private sector alternatives to regulation for risk management.

George Eads has been a member of the President's Council of Economic Advisors, where he oversaw activities concerned with regulatory reform and chaired the Regulatory Analysis Review Group. He is currently dean of the School of Public Affairs at the University of Maryland.

Baruch Fischhoff is an experimental psychologist at Decision Research, who has done extensive research on the psychology of decision making and risk assessment. He is coeditor of *Acceptable Risks*.

Robert M. Friedman is senior analyst at the Office of Technology Assessment. He directed an assessment of transported air pollutants for the congressional committees responsible for the reauthorization of the Clean Air Act.

Mary Gibson is associate professor of philosophy at Rutgers University. She is the author of the recent book *Workers' Rights*.

Mark MacCarthy is professional staff member on the Energy and Commerce Committee of the U.S. House of Representatives. Formerly he spent three years as an economist with the Occupational Safety and Health Administration.

Laura Punnett has taught occupational health and safety to health-care professionals and is a coauthor of the book *Our Jobs, Our Health: A Woman's Guide to Occupational Health and Safety*.

Peter Railton is associate professor of philosophy at the University of Michigan. He writes on the philosophy of science, on social and political philosophy, and on the problems that arise at the intersection of these two areas.

James L. Repace, environmental policy analyst at the Environmental Protection Agency, has published numerous articles on problems relating to indoor and outdoor air pollution, particularly on the risks of passive smoking.

Clifford S. Russell, an economist, is senior fellow and director of the Quality of the Environment Division at Resources for the Future. His publications include articles on effluent charges and the trading of pollution permits.

Samuel Scheffler, associate professor in the Department of Philosophy at the University of California at Berkeley, is the author of *The Rejection of Consequentialism.*

Judith J. Thomson is professor in the Department of Philosophy and Linguistics at the Massachusetts Institute of Technology. She is the author of many articles on topics in moral theory. Her books include *Acts and Other Events* and *Self-Defense and Risks.*

Langdon Winner is associate professor of politics and technology at the University of California at Santa Cruz. He is the author of *Autonomous Technology* and many articles on politics, technology, and social change.